Table of Contents

Acknowledgements   iv

Preface for Teachers   v

To the Student   x

Unit 1:   Introduction to Machine Tool Technology   1

Unit 2:   Shop Safety   7

Unit 3:   Mechanical Hardware   19

Unit 4:   Shop Drawings   27

Unit 5:   Hand Tools   41

Unit 6:   The Language of Mathematics   51

Unit 7:   Measuring Instruments   69

Unit 8:   Layout Tools and Procedures   85

Unit 9:   Sawing Machines   95

Unit 10:   Drilling Machines and Drilling   103

Unit 11:   Turning Machines   119

Unit 12:   Vertical Milling Machines   143

Unit 13:   Horizontal Milling Machines   163

Unit 14:   Grinding Machines   173

Unit 15:   CNC Machining   189

Answers to Unit Questions   199

Study Guide Glossary   209

Index   219

iii

Acknowledgements

The author is indebted to many people for the development of this work.

• Dr. Jon Krug, Dean of Vocational Education at Grays Harbor College, Aberdeen, WA, introduced me to English as a Second Language students enrolled in vocational training programs; he helped my early efforts to teach them technical English for class and work. Jon died in January 2000; he was a true friend to his colleagues and to the students he served. This book is gratefully dedicated to his memory.

• I wish to acknowledge with gratitude several persons within the San Diego Community College District, who have helped me in the past to prepare for the writing of this present book: Mike Buerger, machine tool technology instructor; Sandi Linn, Vocational ESL classroom instructor; Jan Jarrell and Ann Marie Damrau who coordinated previous VESL writing projects in the District.

• At Prentice Hall, Ed Francis, Executive Editor, Education, Career & Technology, has guided and encouraged the writing of this study guide; he combines a vision of meeting the needs of American workers for whom English is not the first language with the practical demands of publishing a book. The author is very grateful for his patience, practicality, and availability when problems arose. Thanks also to Christine Buckendahl, Production Editor, and Joy Dickerson, Copy Editor, who have brought this work to its final form.

• I am also grateful to Richard Kibbe, author of the Prentice Hall text, *Machine Tool Practices*, for which this study guide is intended as a companion volume. I admire immensely the wealth of knowledge about the machining trade that enriches the pages of his text. My thanks go to Mr. Kibbe for his encouragement and help during this writing process.

• Several other readers of the drafts of this study guide added their suggestions and comments; to them I am also grateful.

• For the use of pictures for illustrating this book, many thanks to the San Diego Community College District, to the Clausing Corporation of Kalamazoo, MI, and to the Kibbe text.

• I thank the following persons who lent their voices for the recording of the tapes that accompany this book: Roger Boufford, Bill Coles, Jerri Diaz, and Rich Miro.

• I am indebted to the many students who have worked in my vocational English classes; they are the best barometers of what is useful in acquiring technical English. I admire their hard work, and I rejoice when I see many of them entering the new American workforce.

• Finally, I thank my dear wife, Joanne, for her love and patience during this process.

Pete Stafford, Ed.D.
June 2000

E.S.L. Study Guide for

Machine Tool Practices

Pete Stafford, Ed.D.
San Diego Community College District

Upper Saddle River, New Jersey
Columbus, Ohio

Library of Congress Cataloging-in-Publication Data

Stafford, Pete.
    E.S.L. study guide for Machine tool practices / Pete Stafford.
       p. cm.
    Companion text to Machine tool practices, 6th ed., c1999.
    ISBN 0-13-012354-4
    1. Machine tools. 2. Machine-shop practice. I. Machine tool practices. II. Title.

TJ1185 .S74 2001
621.9'02—dc21
                                                       00-037490

Vice President and Publisher: Dave Garza
Editor in Chief: Stephen Helba
Executive Editor: Ed Francis
Production Editor: Christine M. Buckendahl
Design Coordinator: Robin G. Chukes
Cover photo: SuperStock
Production Manager: Brian Fox
Marketing Manager: Jamie Van Voorhis

This book was printed and bound by Victor Graphics. The cover was printed by Victor Graphics.

10 9 8 7 6 5 4 3 2 1

ISBN: 0-13-012354-4

Preface for Teachers

1. Background to the Writing of this Book:

• An increased number of men and women entering the American workforce are limited in their proficient use of English in their daily lives, in classrooms, and on the job. These workers have come from many lands seeking a better life based on useful work. In school they are known as Limited English Proficient (LEP) students. Recently the term Vocational English as a Second Language (VESL) has been used to describe language activities designed for LEP students who are preparing for a trade.

• In recognition of this influx of new workers into the machine tool trade, Prentice Hall offers this book, *ESL Study Guide for Machine Tool Practices*, as a companion volume to the textbook, *Machine Tool Practices*. The *Study Guide* was written with this major question in mind: *What English language items does a machine technology student need to understand, say, read and write, in order to work effectively in the classroom and the shop?* The purpose of the guide is to assist LEP machine shop students to improve their job-related English skills.

2. For Whom the *Study Guide* Is Intended:

• **Primary audience:**

a. Limited English Proficient (LEP) students for whom English is a second language and who are taking vocational classes in machine tool technology at community colleges, vocational schools, proprietary schools, and community-based training programs.

b. The teachers of machine tool technology classes who encounter the problems of limited English proficiency and who want a resource that they can recommend to LEP students who have such language problems.

• **Other audiences:**

c. English as a Second Language teachers, trainers and others who teach VESL classes, and their LEP students. Classes usually follow either of two models: (1) Up-front VESL classes, offered as a separate language program to be completed before enrollment in the machine tool technology classes; (2) Concurrent VESL classes, taken along with the technical classes. Some programs combine the two approaches by teaching a strong VESL component before technical training and then continuing VESL as the training progresses.

d. LEP students who are already working for a company in the machine tool trade and who wish to upgrade their skills. Some trainees are hired before they finish classes, but may wish to continue their studies at night and in off hours. Such students will be using the textbook, but may also want to work on their trade-related English through the use of the *Study Guide*.

e. Although the focus of the *Study Guide* is on the LEP student, some Adult Basic Education students who are taking vocational classes in machine tool technology may find the extra practice and the reading aids helpful, especially if they are experiencing difficulties with literacy tasks. The audiotapes, the illustrations, and the nomenclature exercises may be of particular interest.

3. The Main Features of the *Study Guide*:

• To develop English skills useful in the machine tool trade, the *Study Guide* offers a series of language activities, contained in fifteen units. Each activity is numbered sequentially within a given unit, using a unit number and an activity number; for example, 2-01, 2-02, 2-03, etc. This allows easy location of activities when they are referenced elsewhere. The fifteen units are in the same order as their parallels in the textbook.

• Here are the main features of the *Study Guide*:

a. Unit Objectives:

   The *Study Guide* lists language objectives at the beginning of each unit. These objectives support the learning objectives of the textbook.

b. Vocabulary Activities:

   The *Study Guide* lists vocabulary words for each unit; these words are taken from the textbook and the workbook. Each word is defined and is used in a sample sentence. The words appear again in the readings that follow each list. The pronunciation of vocabulary is given on an accompanying tape.

   The words are of four kinds:
   1) machine nomenclature, e.g., *spindle, arbor, collet*
   2) other technical words from the trade, e.g., *to broach, orthographic projection, chamfer*
   3) ordinary words used with a technical meaning, e.g., *to turn, drawing, pocket*
   4) important context words for technical English, e.g., *visualize, intersect, stationary*.

   The study of these vocabulary items will better equip LEP students to read key passages in the textbook and the workbook.

c. Reading Activities:

   The *Study Guide* uses readings written in the range of sixth to eighth grade level to convey key ideas from the textbook which are written at the college level.

   1) Some passages simplify important ideas from the textbook.

   2) Some passages expand ideas for which an ESL student may need more explanation.

   3) The passages represent the range of materials covered in the textbook: procedures, descriptions, information about operations and accessories, and special attention to learning the names and purposes of important machines and tools.

   4) Important reading techniques are explained; practice in their use is provided by sending the student to key passages in the textbook. The *Study Guide* is most beneficial when it is used as a tool for getting more from the textbook.

   5) Comprehension checks are given for the reading passages. These tests help the LEP student to practice the style and form of questions asked in the textbook's unit tests.

   6) Readings are illustrated by the author's drawings or by illustrations from the textbook.

Main Features (continued)

d. Audiotapes:

Accompanying the *Study Guide* are two 90-minute audiotapes. The tapes are to assist the LEP student to understand and converse in the spoken English of the trade. There are several situations where this is important: first, in the classroom, the student listens to lectures and demonstrations, participates in discussions, asks questions, and answers the questions of the teacher; second, in the machine shop itself, the student must hear and understand instructions from the teacher on the use of the machines. The student needs to be able to ask questions about carrying out projects and about the operation of the machines. The student needs to be able to understand verbal instructions about safety.

To achieve these listening, understanding, and speaking goals, the *Study Guide* provides several kinds of tape activities:

1) "Machine Nomenclature" uses a set of three pictures for each new machine, tool group, or machine accessory to teach the names and functions of important parts:

   a) Presentation page: a clear line-drawing depicts the important features of a machine, a group of tools, or accessories. The features are labeled by name; the names are pronounced on the audiotape, with pauses for repetition.

   b) Matching page: the picture is presented again with accompanying letters which are matched with numbered names.

   c) Audio Quiz page: The student looks at a third picture of the parts, listens to the names of the parts, read in a different order, and then locates the parts on the picture.

   Secondary uses of the Audio Quiz page are:
   (1) Purpose Quiz: the student is asked to listen to a list of numbered uses and match those with the pictures on the page.

   (2) Writing Quiz: The student looks at the pictures on the Audio Quiz page and writes the names.

2) "Shop Talk" is a series of audio exercises that requires students to recognize and produce speech which is typically used in a machine shop environment:

   a) Some exercises have the student interpret meaning, as in warnings, directions, suggestions and statements from a teacher, a supervisor, a fellow student, or a co-worker.

   b) Other exercises have the student produce speech as in asking questions and giving answers.

3) "Conversations" is a series of recorded dialogues that require careful listening and repetition. A cast of characters puts a human face on some of the shop activities.

Main Features (continued)

e. Writing Activities:

Writing is an important language skill used in the classroom and the shop. Students need to be able to take notes during lectures and demonstrations, as in other high school or college classes in which large amounts of information must be learned. Getting a job, retaining a job, and advancing in it are linked, in part, with the ability to write. On the job, workers often need to fill out forms, write reports, and make notes for the next shift. The *Study Guide* includes activities like these:

1) The student writes individual words to fill in blank spaces by finding the missing words in a word bank, or by listening to dictated sentences.

2) The student practices writing short answers to open-ended, written questions, like those given in the textbook.

3) The student listens to the tapes and writes brief notes, e.g., while listening to a short classroom lecture.

4) The student practices the use of standard abbreviations for machine shop writing tasks.

f. Activities Using the Textbook:

The textbook, *Machine Tool Practices*, is the main printed resource for the student to learn the information necessary to become a machinist. The activities in the *Study Guide* will assist the student to get more from the textbook. The activities are to supplement and not to supplant the use of the textbook. The *Study Guide* offers these activities:

1) Summarizing and outlining important information from the textbook while preserving the content of the text.

2) Exercises in the use of the textbook's *Glossary*, as a source of vocabulary.

3) Practice in locating information in the various tables found in the Appendices of the textbook.

g. Practice for Assessment Tests:

The *Study Guide* provides practice with the vocabulary and technical background needed to answer test questions given at the end of the textbook's units and in the Instructor's Manual. The *Study Guide* gives sample tests for practice.

h. Illustrations:

Clear illustrations help students to visualize and identify key pieces of equipment and key ideas needed by machine tool workers. Some of the illustrations are only sketches to aid students in language production; some are not made to strict drawing standards.

Main Features (continued)

The *Study Guide* uses some photos and drawings from the textbook.
The *Study Guide* references other illustrations in the textbook, but does not reprint them.
The *Study Guide* provides many new drawings and pictures which help to illustrate
   vocabulary, shop equipment, major machines, and important ideas.
For parts identification, the *Study Guide* uses clear line-drawings.
Some illustrations are used with language practice, as in the nomenclature exercises.
An icon of a tape casette is used as a visual indicator for sections of the text which can be
   found on the tape. ESL students are accustomed to seeing these visual cues.

4. English Language Skills Needed by Students in a Machine Tool Environment:

Here is a list of language skills needed by students in the machine shop; the *Study Guide* seeks to improve student functioning in these areas:

a. Listening to and Understanding English:
1. understand spoken words, phrases, and sentences that occur in the machine shop, including warnings, suggestions, instructions, comments and descriptions.
2. understand material spoken at lectures and demonstrations.
3. understand directions given by teachers in class and by supervisors on the job.
4. understand questions about machine shop problems, practices, and procedures when asked by shop staff, classmates or co-workers.

b. Speaking English:
1. produce shop-related speech which is clear, understandable and appropriate to the work environment, including technical terms and nomenclature.
2. ask questions about, or otherwise clarify, missing or incomplete information.
3. ask questions about problems, or explain a problem to someone else.
4. make helpful suggestions about class or work.
5. start and carry on conversations with co-workers during break times.
6. read important information out loud so it can be understood.

c. Reading English:
1. comprehend these categories of words in the textbook and workbook:
   a) technical terms, including the nomenclature of the machines.
   b) ordinary words with special technical meaning in a shop context.
   c) ordinary words and structures that form the larger context.
2. scan the textbook for needed information.
3. skim the textbook in order to summarize the gist of what is being said.
4. read and answer questions on classroom examinations.
5. use text tools like the table of contents, index, and tables to locate information.
6. read charts, tables, engineering drawings, picture labels, and standard symbols.

d. Writing English:
1. write complete sentences about shop-related matters with attention to punctuation, spelling, and grammar.
2. write a report on a shop problem such as safety or production.
3. write a memo making a suggestion to a supervisor or to co-workers.
4. use standard abbreviations.
5. make sketches or use pictorial means to convey ideas or problems.

To the Student

Congratulations to you for choosing the manufacturing industry as a place to work. Many students have gone before you and are working today, earning good salaries, and making useful products for people everywhere. This book, *ESL Study Guide for Machine Tool Practices*, is intended for any student who needs help in reading, writing, listening and speaking the English that will be needed for jobs in the machining trade.

1. What this book can do for you:

Build your vocabulary.
New words are introduced in each unit of this book in the *Vocabulary Lists*. Most of the words are the ones that machinists use to talk about their work and about the machines. You will learn the words by seeing them on paper, hearing them on *audiotapes*, and using them in writing and speaking. A picture (called a *logo*) of an audiocassette is a visual sign that you should play the next section of the audio tape.   logo

The *Study Guide* has its own dictionary of words at the back of this book (the *Glossary*); you can look up a word and see how it is used in the machining workplace.

Practice communicating in the shop.
You can practice sample *conversations* on important machine shop topics by using the tapes and working with a study partner. You need to be able to: (a) ask questions about your job and the equipment; (b) explain problems to other workers and to teachers or supervisors; and (c) to understand directions and be able to give them to others. The parts of this book called *Shop Talk* will help you with these important language skills.

Help you understand the textbook.
The textbook, *Machine Tool Practices*, has a rich supply of knowledge about everything you need to know to be a succesful machine operator. The *Study Guide* has picked out important parts of the textbook and written shorter *readings* that are at a level which will be easier to read. The *Study Guide* uses many *line drawings* to show you important parts of the machines and teach you their names.

Practice taking tests.
The *Study Guide* gives you *comprehension tests* on all the readings you do to see if you have understood the most important ideas.

2. How to use this book:

Let your teacher and the textbook lead you.
The units in the *Study Guide* are in the same order as the sections in the textbook. The *Study Guide* is not a substitute for the text; it is an extra help for you in vocabulary and in written and spoken English.

Each class day, find out from your teacher where you are to study in the textbook, in the workbook, and in any handouts. Then find the parts in the *Study Guide* that go with those materials. When you know what ideas will be covered, you may want to look at the *Study Guide* first, to warm up on easier ideas and pictures. Then you may be better able to work with the more difficult ideas and words in the textbook.

Go at your own speed.
You are an individual student; you may be different from some other students in how fast or slow you understand information. Don't compare yourself with others. Use the time that you need to learn.

Listen to the tapes.
Words and expressions enter your memory and are ready for your use through repeated practice. Listen to the words, the questions, and the explanations on the tape over and over. Then say those words and phrases again and again. Compare your pronunciation to the voices on the tape. Find English speakers who will listen to you and help you with the words.

Do the exercises.
Read and reread the readings in the *Study Guide*. Then do the exercises and other activities. You need to work hard at these activities. Then go to the textbook and read the sections. You will need to read certain parts of the text several times. You may want to write key ideas down as you read.

Keep reviewing.
It is important to review what you learn. Go back over lesssons you have already studied. Listen again to earlier parts of the tapes. Look at the pictures. Test yourself on the names of important parts by covering up answers and naming what you see.

Keep a vocabulary notebook.
You will be able to ask better questions and give better answers as you learn more words in English. It is a good idea to keep a small notebook that will fit into your pocket. In this notebook, you can write down words that you hear or see during class, in conversation, or in reading. You can make your own vocabulary lists to learn.

Find a time and place for study.
It is important to study, especially when you are working on language skills. At the beginning of each week, look ahead and see what you will be doing during the week. Schedule times for study during the week. If you do not have a quiet place at home to study, find a place at the library or come to the classroom early.

Practice your English outside of class.
The more you are able to practice your English, the more easily you will be able to use it in class and at work. It is sometimes easy to sit quietly in class and never try speaking. It is also easy to only speak your native language at home and at other times outside of class. You may want to find an English-speaking person at class or work who is willing to help you with your English conversation and pronunciation. Other members of the class who are from your language group can help you. Get to know the teacher; talk with him or her outside of the teaching time. Watching English-speaking TV and listening to the radio can also be helpful.

Enjoy this book and your new work. You have the ability to learn the English you need for this job. Good luck to you in the future.

Unit 1: *INTRODUCTION TO MACHINE TOOL TECHNOLOGY*

ASSIGNMENT: **Read, study, and complete pages 1 to 6 of this book. Then read pages v and vi and pages 1 to 5 in the textbook,** *Machine Tool Practices, Preface and Introduction.*

OBJECTIVES for this unit:
You should be able to:
1. Tell what a machinist does in his or her work.
2. Write correctly some vocabulary words that are useful as you begin your study.
3. Converse about different jobs in the machine tool trade.

1-01: VOCABULARY LIST A
Directions: Study the vocabulary. Write the missing words in the blank spaces.

1. machinist   a worker who uses large machine tools to make parts out of materials such as metal. *Example:* I want to be a _____ , because the pay is good and the work is interesting.

2. to remove   to take away. *Example:* The machinist is able _____ metal by cutting, drilling, sawing and other steps.

3. rough stock   a piece of unworked metal, like steel, from which the machinist will make a part. *Example:* Jose cut off a 9-inch piece of _____.

4. workpiece   the part as it is being worked on by cutting, drilling, grinding and other machining steps. *Example:* It is good to measure the _____ as you work on it.

5. chips   small pieces of metal that are cut away from the workpiece. *Example:* Don't clean up _____ with your hands!

workpiece

chips

6. trade   a kind of work that requires special skills. *Example:* If you want to succeed in the machine tool _____, you must learn many things.

7. to carry out   to do what you planned to do. *Example:* Marie was able _____ _____ the directions given her by her teacher.

8. CNC   in *computer numerical control* a computer tells the machine what to do to make a part. *Example:* Beatriz learned how to use a _____ machine to make beautiful parts for sailing boats.

9. to program   to give a computer some commands for doing something. *Example:* Ly was able _____ the CNC machine to make many auto parts.

10. core   the most important ideas and skills at the heart of an education program. *Example:* You should learn the skills that are part of the _____.

1-02: READING
Directions: Read the following paragraphs.

What Does a Machinist Do?

A machinist is a worker who makes useful parts for our modern machines; these parts are used in automobiles, planes, stoves, vacuum cleaners and other manufactured products. The machinist operates large machines that make these parts from a rough piece of metal by removing material. Large machine tools can saw, drill, mill, turn, and grind the metal until the part is finished.

The material with which the machinist starts is called **rough stock**. While the material is being worked on, it is called a **workpiece**. The final product is called a **finished part**. These are the usual steps for making a part:

1. Cut off a piece of rough stock of the correct length.

2. Begin to remove material from the workpiece; the pieces of metal are called **chips**.

3. Complete the part; put a finish on it; check it.

rough stock

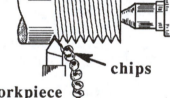
chips
workpiece

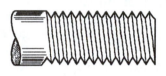
finished part

These steps have been a part of machining since the beginning of the trade. However, it is good for you to know that the way these steps are carried out has changed over the last ten to fifteen years and that more changes can be expected. One of the main things that has changed is the use of computers to control how the steps are done; a computer can be programmed to guide the cutting tools into the workpieces with no mistakes. A computer program directs the tools in these steps in a process called **computer numerical control**, known as **CNC** machining. With this kind of machining, a machine operator can produce a part over and over in large numbers.

Because of the changes in how parts are made, the kinds of jobs in this trade have also changed; for example, formerly most of the work was done by a machinist who knew how to run all of the big machines; now much of the work is done by machine operators who may have a more limited knowledge of just one of the machines. However, the basic processes shown above are the same, even in CNC machining. And so today's machine tool student should learn a core of knowledge that will still be very valuable in understanding and doing machining jobs, no matter what changes come in the future. The authors of the textbook give you this training as you work through the textbook and do the projects in the workbook and the assignments of your teacher.

1-03: EXERCISE
*Directions: Read the questions. Circle the **best** answer.*

1. What is the main activity a machinist does on the job?
   a. drives cars or flies planes
   b. sells manufactured products
   c. makes parts by removing material
   d. all of these

2

2. Which of these jobs can large machine tools do?
   a. saw pieces of metal
   b. drill holes in the metal
   c. cut and grind metal parts
   d. all of these

3. Which is the first step in making a part?
   a. put a finish on the part
   b. cut, drill, or grind a workpiece
   c. cut off a piece of rough stock
   d. none of these

4. What is the name for this picture which shows the part after its final check?
   a. workpiece
   b. finished part
   c. rough stock
   d. name not given

5. What is the name for this picture which shows the part while it is being worked on?
   a. workpiece
   b. finished part
   c. rough stock
   d. name not given

6. Which of these statements is *not* true?
   a. There has been very little change in how parts are made in the last twenty years.
   b. There are more and more changes taking place in how parts are made.
   c. Today computers are more widely used in the making of parts than ever before.
   d. Because of changes in manufacturing parts, job duties are changing.

7. What do the initials CNC stand for?
   a. computerized nominal control
   b. computer numerical control
   c. controlled nominal computers
   d. computer national company

1-04:   "SHOP TALK"   *QUESTIONS & ANSWERS*

> • *Use the "pause" button on your cassette recorder when you want to stop for a short time.*
> • *Use the "rewind" button when you want to repeat something.*

Directions: Listen to these questions on the tape. Practice your pronunciation as you repeat each question. Listen to the answer for each question. Fill in the blanks with what you hear.

Student: Where do you keep the rough stock in this shop?
Teacher: It's in the stock _____. Go and tell Carlos what you _____.

Student: Is the machine tool trade changing?
Teacher: Yes, but learn the _____ ideas and skills and you'll do _____.

Student: Does a machine operator know as much as a machinist?
Teacher: No, but an _____ can become a _____, with more training.

3

1-05: VOCABULARY LIST B
Directions: *Study the vocabulary. Write the missing words in the blank spaces.*

1. career   a work that a person trains for and continues doing for a lifetime.
   Example: Congratulations for making machining your _____!

2. unrelated   not connected to, not leading to anything else. *Example*: Some people work at several _____ jobs, instead of a career.

3. entry-level job   a first job that requires less skill, but is a way of getting started on a career.
   Example: I saw an ad in the newspaper for an _____.

4. to specialize   to develop special knowledge and skills in only one part of a career, in order to do the job very well. *Example*: Joe wants _____ in working with computers.

5. field   any general area of knowledge and work which requires special training.
   Example: Tran has decided to enter the _____ of teaching.

6. conventional   the ordinary way of doing something as practiced by those who came before. *Example*: It will be good to learn _____ machining before you learn how computers can help.

7. accuracy   the condition of being correct, especially in getting correct measurements.
   Example: Many airplane parts require a high amount of _____.

8. to calculate   to compute; to use mathematics to get an answer. *Example*: Leo is trying _____ the length of the part with accuracy.

9. production   the act of making something like parts, especially in large quantities.
   Example: Alberto works in a _____ factory that completes an average of 5,000 parts each day.

10. efficiently   doing something well without waste, delay, or extra expense. *Example*: Victor tried to produce many parts accurately and _____.

1-06: READING
Directions: *Read the following paragraphs.*

CAREERS AND UNRELATED JOBS

There is a difference between a career and working at many different jobs. Someone who has a career enters a field and stays with it for a large part of his or her life. During that time the person learns more and becomes very skillful at work. Other persons choose to work at many different jobs during their lifetimes; they may work for a year or two at one thing and then go to something completely different. These persons have a series of unrelated jobs, not a career.

> **A Machining Career**
> 4. Supervisor
> 3. Machinist
> 2. Operator
> 1. Entry level job

A person who wants a career can receive training before starting to work. After training, the new worker usually takes an entry-level job at lower pay. The person may continue going to school at night in order to learn more. With years of experience, this worker will make more money and can specialize in an area within the field. Some career workers can also supervise other workers, if they are good working with people.

1-07: EXERCISE
Directions: Read this story about two workers; then answer the two questions.

Ramon came to the United States at the age of 20. He worked for a plant nurs-ery during the day and went to English classes at night. He entered a machine shop training program and got a certificate after two years. He got an entry level job as a machine operator at a factory that made automobile parts. He continued in school at night and was made a general machinist at the factory after three years; after five more years he became a supervisor and is doing that now.

 Betty left high school before graduation to help her dad in the furniture business for three years. She didn't like the work, so she went to work in a flower shop for six months. Then she took some classes at a beauty school and worked as a beautician for two years. Lately she has been working as a waitress.

1. Who has a series of unrelated jobs? _____   2. Who has a career? _____

1-08: READING
Directions: Read the following paragraphs.

CAREERS IN THE MACHINE TOOL TRADE

Machine tool technology is a field that has many different career paths. You have now started your training to enter this field of work. It is an exciting field with many changes going on and many career opportunities for those who have talent, who work hard, and who continue to learn.

1. Machine Operator:
The tasks of the machine operator include:
  a. Operating CNC controlled machines.
  b. Loading material for parts into the machines and inspecting the parts as they are made.
  c. Watching to see that the machines run well and changing the cutting tools when needed.

To become a machine operator, a worker must be able to:
  a. Understand how the large conventional machine tools work.
  b. Read the drawings which show how parts are to be made.
  c. Understand cutting tools and choose the correct ones for the job.
  d. Measure the sizes of parts.
  e. Know the mathematics that applies to making parts.

2. General Machinist:
The tasks of the general machinist include:
  a. Reading shop drawings and deciding how to set up equipment for making the parts.
  b. Calculating the sizes on the parts and correct settings on the machines.
  c. Operating all of the large conventional machine tools like drills, lathes, and mills.
  d. Measuring the sizes of parts for accuracy and making other quality checks.

The general machinist will often work in places called job shops. A job shop is usually a smaller business that gets orders for one part or a small number of parts for special situations; for example, a boat owner needs a special pulley for raising a sail. The jobs that come in are often quite different and sometimes require all of the wider experience and knowledge of the general machinist.

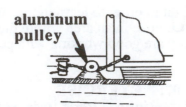
aluminum pulley

The machine operator is more likely to work in a larger production factory where hundreds and thousands of the same parts are being made for large markets; the computer-controlled machines can produce these large numbers of parts accurately and efficiently.

1-09: EXERCISE   *Directions: Match the letters with the numbers.*

_____ 1. job shop

_____ 2. production factory

_____ 3. general machinist

_____ 4. machine operator

_____ 5. entry-level job

_____ 6. career

a. can use all the conventional machines to make parts.

b. work which you train for and continue during your life.

c. can use computer-controlled machines to make parts.

d. a large business that makes large numbers of like parts.

e. a business that makes smaller orders of parts.

f. work with which to begin your life of work.

Note: Read pages 3 to 5 in your textbook; write down the job duties of three other workers in the machine tool field.

    1. _____

    2. _____

    3. _____

1-10: "SHOP TALK"   QUESTIONS and ANSWERS on CAREERS

Directions: Listen to the student's questions and the counselor's answers on the tape. Fill in the blanks with what you hear.

Student :   What skills do I need to succeed as a machinist?

Counselor: You need to be able to _____ pretty well; there are many things to learn as you read. You need to know some basic _____; if you can add, _____, multiply, and _____, the class will teach you the rest and how to _____ things accurately. You need to be able to stand on your _____ at a machine while you're working. It helps a lot if you've had some experience using _____ before and like working with them. Like any job, you need to get along with other _____.

Student: What are the rewards and satisfactions of this kind of work?

Counselor: The _____ are good in machining; you'll start well above minimum _____ and in a few years you can make good _____. You'll have many opportunities to advance in your _____. You'll have a sense of pride as you make useful and beautiful _____.

6

Unit 2: *SHOP SAFETY*

ASSIGNMENT: Read, study and complete pages 7 to 18 of this book. Then read pages 5 to 12 in the textbook, *Machine Tool Practices, SECTION A, Unit 1, Shop Safety.*

OBJECTIVES for this unit:
You should be able to:
1. Identify common shop hazards.
2. Identify common shop safety equipment.
3. Identify key safety practices for the machine shop.
4. Converse about shop safety hazards, shop safety equipment, and shop safety practices.

2-01: VOCABULARY LIST A
Directions: Study the vocabulary. Write the missing words in the blank spaces.

1. safety

being free from danger, injury, or damage.
Example: We are studying _____ in this unit.

2. hazard

a danger.
Example: Oil on the shop floor is a _____ .

3. accident

being hurt without wanting to be hurt.
Example: Jose had an _____ while he was talking with Bill.

4. injury

getting hurt.
Example: If you have a bad _____, you may not be able to work.

5. permanent

lasting, a situation that will not change. *Example:* Some injuries are _____.

6. handicapped

being limited in what you can do because of permanent injury. *Example:* A person who is _____ may need special equipment.

THINK ABOUT SAFETY

Learn an Attitude of Safety!

7. absent

not be at work, missing from work.
Example: Bill was _____ from work for three weeks.

8. attitude

a mental outlook on some situation.
Example: Good safety habits come from a good _____ .

9. practice

an action which is learned and repeated many times, a habit.
Example: Putting safety glasses on is a good _____.

10. to protect

to keep from injury.
Example: Use safety glasses _____ your eyes.

7

2-02: READING
Directions: *Read the following paragraphs.*

Safety Is Important

Safety is important for you, your employer, and your family. Each year many workers in the USA, including some in the machine shop, have accidents on the job. When you have an injury, you are sometimes absent from work. You can even become handicapped without the full use of your eyes, your hands, or other parts of your body. Your employer loses your good work and the money spent on your training. The company has to hire and train other workers. Your family loses money and has to take care of you until you are well. Sometimes workers are injured permanently and are not able to return to their jobs.

This unit will look at the hazards in the machine shop. The unit will also identify safety equipment and safety practices which will protect you and those working around you. The most important practice is to learn an attitude of safety. This is the habit of always thinking about safety and trying to work in a safe way.

In later units you will continue to learn safety practices for particular machines like the lathe, the milling machine, and the drill press.

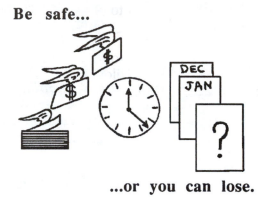

Be safe...

...or you can lose.

2-03: EXERCISE
Directions: *Read the questions. Circle the **best** answer.*

1. Who loses when you are hurt at your job?
   a. your family
   b. your employer
   c. yourself
   d. all of these

2. Which of these is a handicap for the worker?
   a. losing time on the job
   b. losing money from the job
   c. losing the use of a hand or an eye
   d. losing the employee's work

3. Which word is closest in meaning to the word *accident*?
   a. injury
   b. handicapped
   c. hazard
   d. attitude

4. Which of these is *not* studied in this unit?
   a. common shop hazards
   b. milling machine safety
   c. safety equipment
   d. an attitude of safety

5. Which of these shows an attitude of safety?
   a. care when using shop equipment
   b. thinking about safety rules
   c. reading safety signs
   d. all of these

8

2-04: VOCABULARY LIST B
Directions: Study the vocabulary. Write the missing words in the blank spaces.

1. potential   possible, capable of happening. *Example:* We will study the _____ dangers of the machine shop.

2. to rotate   to turn around a center point or axis, to spin, to revolve. *Example:* Cutting tools are designed _____ at very high speeds.

3. to secure   to hold something tightly in place. *Example:* Before you cut the workpiece, be sure _____ it to the milling table.

4. to warn   to let someone know about possible danger. *Example:* The teacher needs _____ all the students about hazards in the shop.

5. to pinch   to squeeze between two surfaces. *Example:* If you get your hands too close, it can cause you _____ your fingers in those gears.

6. guard   a piece of metal which covers moving parts on a machine to protect the machine operator. *Example:* Make sure that _____ is on the machine before you turn it on.

2-05: READING
Directions: Read the following paragraphs.

SAFETY HAZARDS AND SAFETY PRACTICES

The machine shop has many potential dangers; you need to be careful there. At first the shop is new to you, so you may be afraid of possible hazards. With experience in the shop, you will work well, without fear, but with respect for the power of the machines and the hazards of shop work. Now learn some shop hazards and some safe habits for working.

     1. HAZARDS TO THE EYES: The large machines which saw, drill, turn, cut and grind have moving parts which rotate at very fast speeds. As each machine operates, it removes material from the workpiece. Metal chips, particles, and dust are produced. Workpieces can come loose from where they were secured and can be thrown out from the machine or across the shop. These are hazards for the eyes. You also need to watch out for chips and flying objects from the machines of your co-workers.

Here are some examples of eye protectors:

safety glasses   safety goggles   face shield

Safety Practices for the Eyes:
a) Use eye protection like safety glasses, safety goggles, or face shields.
b) Warn your co-workers about eye hazards.
c) Listen to eye-safety directions from your instructor.

> • *Use the "pause" button on your cassette recorder when you want to stop for a short time.*
> • *Use the "rewind" button when you want to repeat something.*

Directions: *Listen to these warnings on the tape. Practice your pronunciation as you repeat each warning. If you receive a warning, thank the other person. Fill in the blanks with what you hear.*

Student 1:   I'm going to turn on the machine; please stand back.
Student 2:   Thanks for _____ me.

Student 1:   I'm going to turn on the mill. Watch out for _____.
Student 2:   Thanks for telling me.

Student 1:   Now, I'm going to grind some _____. Watch out for the dust.
Student 2:   _____ for letting me know.

2-07: "SHOP TALK" *QUESTIONS and DIRECTIONS*

Directions: *Listen to the students' questions and the instructor's directions on the tape. Practice your pronunciation as you repeat each question and direction. Fill in the blanks.*

New student: Can I turn on the machine now?
Instructor: First, put your _____ on.

Another new student :   Can I turn on this milling machine?
Instructor: First, get some _____ on.

New Student: Tito wants to help me grind some parts. Is that okay?
Instructor: First, both of you get _____ and put them on.

2-08: READING
Directions: Read the following paragraphs.

2. HAZARDS TO THE HANDS: As the work progresses chips pile up on and around the machine. These chips are very sharp and can cut you if you use your bare hands to brush the chips away. Many of the machines have gears and moving parts which can cut, pinch, or cut off fingers. Cutters, twist drills, toolbits and saw blades can cut through fingers and arms.

Safety Practices:
a) Don't wear gloves except when you are welding.
b) Keep your fingers at a safe distance from moving parts
    when the machines are running.
c) Turn off the machine before you measure a workpiece.
d) Unplug the machine before you make adjustments or measure anything.
e) Don't brush chips away with your hand or a rag; use a brush for that.

2-09: "SHOP TALK"   *STATEMENTS and DIRECTIONS*

Directions: *Listen to the students' statements and the instructor's directions. Fill in the blanks.*

Student 1:   I'm going to clean up these chips.
Instructor: Wait a minute. Don't use your _____. Use a _____ for that.

Student 2:   I'm going to clean up these chips with a _____.
Instructor: Just a moment. Don't use that rag. Use a brush instead.

Student 1:   I want to measure this _____.
Instructor: Just a second. Stop your machine first.
              Then measure the workpiece.

brush

Student 2:   I want to _____ the belts on this drill press.
Instructor: First, stop your _____. Then unplug it,
              before you adjust it.

Student 1:   I'm almost _____ with this part.
Instructor: Let the machine _____ by itself; don't try to slow it with your hand.

2-10:   READING
Directions: Read the following paragraphs.

3. HAZARDS TO THE LUNGS: Sometimes dust is produced during a grinding operation. This dust can be breathed into your lungs and be absorbed by your body. This can be very dangerous to your health. You can also inhale poisonous fumes from metals which are heated during operations like heat treating.

Safety Practices:
a) Use coolant to cut down on dust particles.
b) Use a vacuum dust collector to catch much of the dust.
c) Use respirators with fresh filters to remove dust particles before you breathe the air.
d) Keep work areas well ventilated to blow away dangerous fumes.

2-11: "SHOP TALK"   *STATEMENTS and DIRECTIONS*

Directions: *Listen to the students' statements and the instructor's directions. Fill in the blanks.*

Student 1:   I'm going to _____ these parts.
Instructor: Before you do, make sure the _____ is running and turn on the
              vacuum dust collector.

Student 2:   I'm going to _____ this tool at the grinder.
Instructor: Before you do, get a _____ from the storage locker.

Student 1:   I want to heat-treat this part. Is it _____ to do that?
Instructor: Make sure you have good _____. Turn on the blowers and
              open the window.

2-12: READING
Directions: Read the following paragraphs.

4. HAZARDS TO THE BACK: It is possible to injure your back if you do not lift things correctly. Bending over to pick things up from the floor is not the way to do it. If objects are large or heavy, you can injure your back by trying to lift and carry things by yourself.

Safety Practices:
a) First, decide if this is an object that you can lift by yourself.
b) To lift a low object, bend your knees, keep your back straight, and lift with your legs.
c) To lift and carry heavier objects, get help from a coworker or use a lifting machine like a forklift.

right   **wrong**

(San Diego CCD)

2-13: "SHOP TALK"   *WARNINGS*

Directions: Listen to these warnings. Practice your pronunciation. Thank the other person for the warning. Fill in the blanks.

Instructor: Be careful. That box weighs over _____ pounds.
Student 1:   Thanks for warning me.

Instructor: That's _____ a job for _____ person. You should get some _____.
Student 2:   Thanks for telling me. I'll get some help.

Instructor: That piece of _____ is very _____. Do you need some help?
Student 1:   Thanks for letting me know. I'll get Omar to help me.

2-14: VOCABULARY LIST C
Directions: Study the vocabulary. Write the missing words in the blank spaces.

1. flammable   able to burn, able to catch on fire, combustible.
                    Example: Paper and oily rags are very _____.

2. fire extinguisher   a tool used to put out fire by the use of water or chemicals.
                    Example: Learn the location of every _____ in the shop.

3. housekeeping   the good habit of keeping your tools and materials in order or put away.
                    Example: Frank has tools and materials everywhere. He needs to do some _____ in his work area.

4. compressed air   air which is held under pressure. Its main purpose is to run some machines.
                    Example: Don't use _____ to blow chips away.

5. horseplay   touching another person in a playful, sometimes rough manner.
Example: _____ in the work area is dangerous, because it takes the worker's attention away from potential hazards.

6. vertical   in an up and down direction.
Example: Carry that piece of stock in a _____ position.

2-15: READING
Directions: *Read the following paragraphs.*

5. HAZARDS FROM FIRE: Flammable materials like oily rags and paper can catch on fire. Electrical equipment and wires can cause electrical fires. Chemicals and some liquids can catch on fire.

There are three common classifications of fires:

Class A Fires	*Class B Fires*	*Class C Fires*
paper, plastic, cloth, rubber, wood	oil, solvents, lubricants, other flammable liquids	electrical fires with electrical wiring

Look for these symbols on different fire extinguishers. The symbols tell what kind of fire the extinguisher will put out.

△A   ☐B   ◯C

Safety Practices:
a) To prevent fires, keep flammable materials away from sources of heat.
b) Know the different kinds of fire extinguishers.
c) Know where the fire extinguishers are located in the shop.
d) Oily rags must be kept in a safety can and emptied daily.

2-16: "SHOP TALK"   *WARNINGS*

Directions: *Listen to these warnings. Practice your pronunciation. Fill in the blanks.*

Instructor: There are too many oily _____ lying around We'll have a _____.

Student 1: Thanks for the warning. I'll put these in the rag _____.

Student 2: I smell hot _____. This power cord may be getting too _____.

Student 1: Thanks for telling me. Let's have the _____ check it out.

Instructor: There's too much stuff piled in front of this fire _____.

Student 2: Thanks for warning me. I didn't notice I was _____ the way.

13

2-17: READING
Directions: Read the following paragraphs.

6. HAZARDS FROM CLUTTER: Pieces of rough stock, finished workpieces, hand tools, boxes, and other items can all collect on the floor around your machine, on your workbench, and in the aisles between machines. This is called clutter, and it can be dangerous for yourself and others. You can trip over an electrical cord or a box, or slip on a piece of round stock, or be unable to move out of the way quickly.

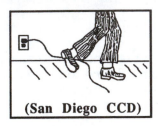

(San Diego CCD)

The Clutter Problem

Safety Practices:
a) Practice good housekeeping.
b) Put away any materials or tools you are not actually using.
b) Keep your area neat and clean.

2-18: "SHOP TALK"   WARNINGS and DIRECTIONS

Directions: Listen to these warnings and directions. Practice your pronunciation. Thank the other person for the warning.

Instructor: Your work area has a lot of _____. That's a safety _____ for you and for anyone walking by. Please _____ it _____ before you do anything else.

Student: Thanks for the warning. I have a _____ with clutter.

Instructor: Take those extra pieces of stock back to the storage area. Put the scrap in the scrap bin. Stack those _____ parts on the tray. And put those _____ tools away.

Student: Thanks. I'll do that _____.

2-19: READING
Directions: Read the following paragraphs.

7. HAZARDS FROM MACHINE TOOLS: You are now studying general safety. As you learn how to operate the individual machines, you will learn that each machine has its own features which can be potentially dangerous. For example, the lathe and some milling machines have open areas near their motors; the machinist must stop the machine and reach in to change the speed by changing belts. For safety, when the machine is running, a guard should be in place, to prevent fingers or clothing getting caught in the moving belts.

Safety Practices:
a) Do not wear jewelry like necklaces, rings, bracelets, wristwatches, nor long-sleeved
   shirts, nor scarves, nor ties, or anything that can get caught in moving machinery.
b) If you have long hair, use a hairnet, or pin it up under a hat.
c) As you learn about each new machine, learn about the safety practices that are required.

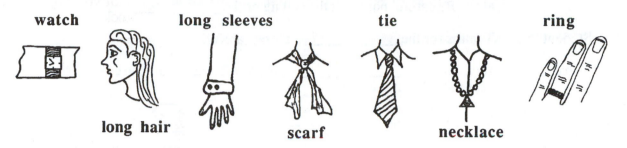

watch   long sleeves   tie   ring

long hair   scarf   necklace

2-20: "SHOP TALK"   *WARNINGS and DIRECTIONS*

Directions: *Listen to these warnings and directions. Practice your pronunciation. Thank the other person. Fill in the blanks.*

Instructor: Students, today I want to warn you not to _____ anything that can get _____ in the machines.

Student 1:   Thanks for the warning. What kind of _____ are you talking about?

Instructor: Don't wear rings, bracelets, necklaces or any other _____.

Student 2:   Thank you! I have a _____ and a _____ on. I'll take them off.

Instructor: Watch out for scarves and ties. Roll your long _____ above your elbows.

Student 1:   Thanks for the warning. I'm also going to pin my _____ up under my hat.

2-21: READING
Directions: *Read the following paragraphs.*

8. OTHER HAZARDS: It is dangerous to get into horseplay with other workers in the work area. It is dangerous to blow away chips and dust with compressed air. It is dangerous to blow compressed air on your skin or clothes. It is dangerous to carry long pieces of material.

Safety Practices:
a) Do not get into "horseplay" in the shop. No playing, wrestling, or scuffling.
b) Do not use compressed air to blow away chips or dust.
c) Do not blow compressed air on your skin or clothing.
d) Carry long pieces in a vertical position or get two people to carry them, one on each end.

2-22: "SHOP TALK" WARNINGS and DIRECTIONS

Directions: **Listen to these warnings and directions. Practice your pronunciation. Thank the person for the warning. Fill in the blanks.**

Instructor: Please take this 8-foot _____ of aluminum back to the _____ area. Be careful not to hit the ceiling or the _____.

Student 2:   Thanks for the _____. I'll be careful.

Instructor: Hey, no _____ in the work area. You can't fool around near those big machines.

Student 1:   Okay, we'll stop. _____.

Instructor: Let me warn you about the compressed _____. You can't use it like a leafblower. It's not for _____ chips off or for cleaning your _____.

Student 2:   Thanks for warning us. I know it's only used for _____ air tools.

2-23: REVIEW of UNIT 2
Directions: **Read the questions. Use words from this list to complete the answers:**

> short-sleeved, jewelry, hands, safety goggles,
> eyes, secure, gloves, safety glasses

1.   Q: What two parts of the body are most important to protect?
      A: Your _____ and your _____ .

2.   Q: What is the primary piece of safety equipment in the machine shop?
      A: Eye protection equipment like _____.

3.   Q: What can I do if I wear prescription glasses?
      A: You can wear _____ over your prescription glasses.

4.   Q: What is proper dress for the machine shop?
      A: Wear _____ shirts, heavy leather shoes, and a shop apron.
         Make long hair _____ with a hair net.

5.   Q: What things are not worn in the machine shop?
      A: No rings, bracelets, necklaces or other _____ .
         For most jobs, no _____ .

Directions: Read the questions. Use words from this list to complete the answers:

> respirator, help, ventilated, knees, ear muffs,
> hands, chips, ventilated, ear plugs, coolants

6. Q: How do I prevent damage to my ears from loud noise?
   A: You should wear _____ or _____ .

7. Q: What can be done to control grinding dust?
   A: During grinding, use liquid _____ and make sure the grinder has a vacuum dust collector.

respirator

8. Q: What else can I do to keep from inhaling dust or toxic fumes?
   A: You can wear an approved _____ . You can make sure the work area is well _____ .

9. Q: How do I protect my back during lifting?
   A: You should bend your _____ , keep your back straight, and lift with your legs. If the load is heavy, you should ask for _____ or use a forklift.

10. Q: What's compressed air for?
    A: It's for operating some machine tools. It's not for other uses like blowing _____ away, drying your _____ , or blowing on your skin or clothes.

Directions: Read the questions. Use words from this list to complete the answers:

> vertical, clutter, extinguisher, hazards, spills

11. Q: Why is housekeeping important?
    A: It's easy to trip and fall, if you leave tools, stock, and other _____ lying around on the floor.

12. Q: What are some good housekeeping procedures?
    A: Clean up oil and other _____ .
       Keep material and stock off the floor. Don't clutter the aisles or the work area.

13. Q: How should long pieces of material be carried?
    A: In the _____ position or with a person on each end.

14. Q: What important questions about safety should you check before you operate a machine?
    A: a. Do I know how to operate this machine?
          b. What are the possible _____ in using this machine?
          c. Are all the guards in place on the machine?
          d. Are all my procedures safe?
          e. Have I made the proper adjustments?
          f. Have I tightened all locking bolts and clamps?
          g. Do I have proper safety equipment? Eye protection?
             Proper clothing? Ear plugs? Respirator? Fire _____ nearby?
          h. Do I know where the stop switch is?
          i. Do I think about safety in everything I do?

2-24: CONVERSATION

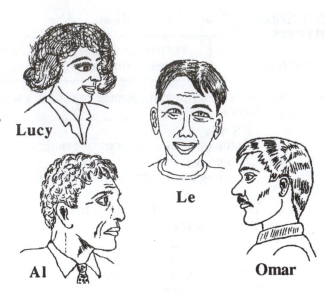

Directions:
Lucy Garcia, Le Tran, and Omar Harris, three machine shop students, are talking with Al Lopez, the machine technology teacher. Listen to the conversation and fill in the missing words. Then rewind and listen to the conversation again. Then get a partner and practice your pronunciation.

Lucy

Le

Al

Omar

Al: Let's talk a few minutes about _____ _____ . Do you know where the fire extinguishers are in this shop?

Lucy: I can see one hanging on the _____ near the toolroom.

Le: Yes, and there's _____ one on the wall near the drill presses.

Al: And there's two _____ : one in the classroom, and one next door.

Omar: I've got a question. How does a fire extinguisher _____ ?

Al: For something to _____ you need fuel, oxygen, and a _____ source. The fire extinguisher usually cuts off the oxygen and reduces the heat.

Lucy: Are all _____ the same?

Al: That's a good question, Lucy. We've got _____ kinds of fires: Class A, which is burning wood, _____ , rubber; Class B, which is _____ liquids, like oils and solvents; and Class _____, which is electrical fires.

Le: So what kind of fire extinguisher do we use?

Al: The fire extinguishers you see here are _____ for all three kinds of fire; they use a chemical foam to _____ _____ oxygen and to cool.

Lucy: Thanks for the fire safety _____ , Al.

Omar: Yes, thank you, Al. I'm learning a lot.

Al: You're welcome!

18

Unit 3: *MECHANICAL HARDWARE*

ASSIGNMENT: **Read, study and complete pages 19 to 26 of this book. Then read pages 13 to 25 in the textbook, *Machine Tool Practices, SECTION A, Unit 2, Mechanical Hardware.***

OBJECTIVES for this unit:
You should be able to:
1. Identify and say correctly the important parts of the unified thread form.
2. Identify the names and abbreviations of the unified thread series.
3. Identify the classes of thread fit.

3-01:   VOCABULARY LIST   A
Directions: Study the vocabulary.   Write the missing words in the blank spaces.

1. fastener

any part which is made to hold other parts together or lock them in place. *Example:*   Each of these is a different kind of _____ :

screw   stud

bolt   nut

2. assembly

a fitting together of parts to make a whole. *Example:* Many parts must be used to make an engine _____ on an automobile.

3. cylinder

a shape which has straight sides and circular ends. *Example:* A soda can is a good example of a _____ .

ROOT BEER

4. continuous

going along without a break or interruption; unbroken; connected. *Example:* A thread is one _____ groove as it winds around the length of the cylinder.

5. round stock

unworked metal, like steel or aluminum, that is cylindrical (shaped like a cylinder). *Example:* Ivan chose a piece of _____ from which he would make a screwdriver.

6. curled

twisted, bent around in the shape of a curl. *Example:* The chip was _____ , like a piece of hair.

curl

7. mating parts

two parts which fit together. *Example:* A bolt screwed into a nut is a good example of _____ .

8. nomenclature

a set of names used in a particular field of work. *Example:* Part of your job is to learn the _____ that is used in the machine shop.

9. diameter

a straight line passing through the center of a circle from one side to the other. *Example:* The major _____ of a bolt is the widest part of the thread.

diameter

10. average

the number you get by adding two or more quantities and dividing by the number of quantities. *Example:* The _____ of 6 and 4 is 5.

19

3-02: READING
Directions: Read the following paragraphs.

ANSWERING SOME BASIC QUESTIONS ABOUT HARDWARE

1. What is mechanical hardware?

The job of a machinist is to make parts which will be used in cars, planes, and other assemblies. Mechanical hardware is the fasteners that are used to hold the parts of an assembly together. One of the machinist's major jobs is to make these fasteners. For example, hundreds and thousands of fasteners are needed in just one airplane; they are what holds the plane together as you lift off the ground.

Examples of mechanical hardware are bolts, screws, nuts, studs, and other fasteners which have threads; washers, pins, and keys are also fasteners. The fasteners are like the buttons and zippers that hold your clothes together.

2. What is a thread? What is a helix?

A thread is a helix-shaped groove that is formed on the outside or inside of a cylinder. A helix is the path of a point which rotates around a cylinder at the same time it moves along the cylinder.

a helix shape 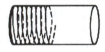 →  **a thread = a helical groove**

In Unit 1 you saw a thread being cut into the sides of a cylindrical workpiece. The thread is one long continuous groove with even spacing as it winds around the length of the cylinder. Threads are often cut into a piece of round stock using a machine called a lathe. The chip that comes away can be long and curly.

threads being cut into round stock

3. How do the threaded fasteners work?

Threaded fasteners have external threads cut into the outside surface of cylinder-shaped bolts, screws and studs, and internal threads cut into the inside of cylindrical holes that are found in nuts and in other parts. The external threads fit into the internal threads and hold the parts of an assembly together. A bolt and a nut are examples of mating threaded pieces.

Below draw a second picture showing the two parts joined by the nut and the bolt.

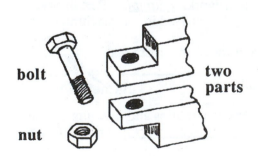

bolt

two parts

nut

3-03: EXERCISE

Directions: For each statement, circle T for "true" or F for "false."

T F 1. One of the machinist's jobs is to make threaded fasteners.

T F 2. A thread is a helical groove cut into the surface of a cylinder or the sides of a hole.

T F 3. The threads on a bolt are called internal threads.

T F 4. All fasteners are threaded.

T F 5. The thread on a bolt will fit into the threads of a nut.

T F 6. Round stock is used to make threaded fasteners on a lathe.

3-04: MACHINE NOMENCLATURE

The Parts of the Unified Thread Form

The unified thread form was agreed to by the United States, Canada, and Great Britain, so that common threaded pieces would be the same in all three countries (standardized). There are other thread forms from other countries and for special purposes. To better understand threaded fasteners, study these parts of the unified thread form.

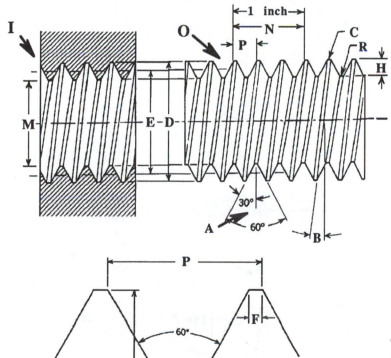

The Names of the Parts:

A   thread angle
I   interior threads
O   exterior threads
D   major diameter
E   pitch diameter
M   minor diameter
N   number of threads per
        inch (*TPI*)
P   pitch
B   helix angle
        or lead angle
C   crest
R   root
H   thread height
        or thread depth
F   crest flat and root flat
        F = 1/8 of P

The picture shows both interior and exterior threads. A diameter is the distance across the width of a circle or cylinder, measured through the center.

The major diameter is the width of the threaded cylinder at its widest part (at the crest). The minor diameter is the width of the cylinder at its narrowest part (at the root). The pitch is the distance from one crest to the next. The pitch diameter is the average of the major and minor diameters.

21

The Parts of the Unified Thread Form

Below are the pictures of the parts of the unified thread form. This begins the steps for learning to pronounce these names and to memorize them so you can say them and write them. There are several "nomenclature" lessons in this book. Learning nomenclature is important, so you can talk clearly with others in the machine shop.

 Directions: *Study this picture as you listen to the pronunciation of these parts of the thread form on Tape 1, side A. Pronounce each part's name after you hear it.*

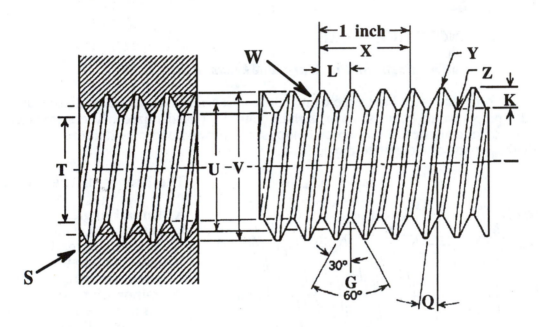

3-06: EXERCISE

Directions: *Look at the pictures. Write the letters next to the names.*

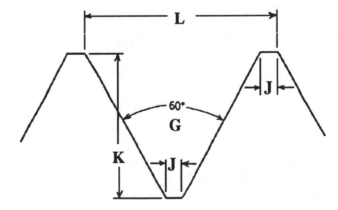

____ 1. root

____ 2. pitch diameter

____ 3. interior threads

____ 4. crest

____ 5. helix angle

____ 6. minor diameter

____ 7. pitch

____ 8. thread height

____ 9. exterior threads

____ 10. thread angle

____ 11. major diameter

____ 12. lead angle

____ 13. root flat

____ 14. crest flat

____ 15. number of threads per inch

22

3-07: VOCABULARY LIST B
Directions: Study the vocabulary. Write the missing words in the blank spaces.

1. to require   to need, to be necessary. *Example:* The machining trade has _____ good eyesight, imagination, and patience

2. to magnify   to make larger. *Example:* When something is very small, to see it clearly you must _____ it. A magnifying glass makes small things look bigger.

magnifying glass

3. series   a group or number of similar things coming one after another. *Example:* There is a _____ of different bolts, arranged by size.

4. coarse   made up of larger sizes. Not fine. *Example:* UNC is an abbreviation which stands for a series of _____ threads; they have fewer, but larger threads per inch than do fine threads.

5. fine   thin, slender, or small. *Example:* UNF (United National _____) is a series of threads that have more, but smaller threads per inch than do the series of coarse threads.

6. tight   p.laced closely together with no space in between. *Example:* The fit of the bolt into the threaded hole was very _____.

7. loose   not tight, having some space between parts. *Example:* The fit of a wing nut over an air filter cover is _____, for easy removal.

8. heading   a name written at the top of the page or paragraph to tell what is below. *Example:* The table on page 14 of the textbook has one _____ that says UNC and another one that says UNF.

9. to classify   to arrange into groups according to some quality particular to a given group. We will want _____ threaded parts according to size, or according to fit, or according to strength.

3-08: EXERCISE
Directions: Practice spelling some important words relating to threads by listening to the spelling and filling in the letters.

1. p __ t __ h

2. r __ __ t

3. __ r __ __ t

4. __ i __ m __ te __

5. t __ __ e __ __

6. __ ng __ __

7. __ __ re __ __   f __ a __

8. h __ __ i __

9. m __ __ or   d __ __ m __ t __ r

10. __ __ t __ ri __ r   t __ __ __ ad __

11. m __ __ or   d __ __ m __ t __ r

12. __ __ t __ ri __ r   t __ __ __ ad __

13. h __ __ g __ t

14. f __ __ t

3-09: READING
Directions: Read the following paragraphs.

CLASSIFYING THREADED PARTS

1. The Size of Threaded Parts

In assembling various mechanical products, such as a truck or a wristwatch, threaded parts are used; a truck may use large bolts, studs, and nuts, and a watch may have many very small screws that require a magnifying glass to be seen and put in place.

How are threaded parts classified? The unified thread form fasteners are often grouped in either the unified coarse series (UNC = **Unified National Coarse**) or the unified fine series (UNF = **Unified National Fine**). Under the heading of UNC or UNF three things are given in a **table** (Look at Table A-3 on page 14 in your textbook): the size, the major diameter, and the threads per inch. Each threaded piece will have a major diameter and a particular number of threads per inch that goes with that diameter. The UNF fine will have more threads per inch than the UNC coarse; for example, the size 1 UNC and size 1 UNF both have a major diameter of .072 inches, but the UNC piece has 64 TPI (threads per inch) and the UNF has 72 TPI.

     a. size: For threaded pieces with a diameter under 1/4 inch, the UNC has the sizes from 1 to 12 and the UNF has the sizes 0 to 12. The fractional inch sizes are given for 1/4 inch and larger. The UNC and UNF series both go from these very small sizes all the way up to a diameter of about 4 inches. Table A-3 on page 14 of the textbook stops at 1 inch.
     b. major diameter: The major diameters are given in thousandths of an inch that are very close to the fractional sizes; for example, 1/4 inch = .248, and 1 inch = .998.
     c. threads per inch: There will be more threads per inch for the fine series when compared with the coarse series: 1 inch in UNC has 8 threads; 1 inch in UNF has 12 threads.

Examples:

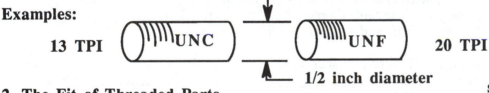

13 TPI   UNC   UNF   20 TPI

1/2 inch diameter

2. The Fit of Threaded Parts

It is possible to make mating parts with tight fits or loose fits. For example, one end of a stud must screw into the threaded hole of an automobile engine block very tightly, and the other end should fit more loosely into the nut that is helping hold on the engine head. Then, when the head nuts are untightened the stud will not come out of its position in the block.

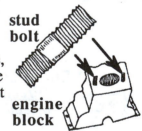

stud bolt

engine block

Unified thread fits are named by these letters and numbers: 1A, 2A, 3A and 1B, 2B, 3B. **Class 1** fits are loose and are used when you want to take something apart easily; **Class 2** fits are used in most common situations; and **Class 3** fits are used when you want a tight fit. **A** is used for external threads, and **B** is used for internal threads.

3. Reading the Thread Classifications

Putting it all together: 1/2 in.-13 UNC 3A bolt would mean: The size of the bolt is 1/2; its major diameter is .498 in.; it is part of the United National Coarse series, and has 13 threads per inch; it is an external thread (A) with a tight fit (class 3).

3-10: EXERCISE

*Directions: Read the questions. Circle the **best** answer.*

1. Which statement is true for a threaded piece which is 1/4 in.-20 UNC 3B?
       a. It has loose fitting exterior threads.   c. It is 1/4 inch long.
       b. It has tight fitting interior threads.   d. It is size 20.

2. Which statement is true for a threaded piece which is 1/4 in.-28 UNF 1A?
       a. It has loose interior threads.   c. It has 28 exterior threads per inch.
       b. It has interior fine threads.   d. None of these.

3. To which number does the abbreviation "TPI" apply in the classification 1/2 in.-20 UNF 2A?
       a. 1/2   c. 2
       b. 20   d. 2A

4. In what situation would you use a Class 3 fit?
       a. For the threads inside a loosely fitting wing nut.
       b. For use with a bolt for most normal purposes.
       c. For the threaded end of a stud bolt which sticks out of an engine block.
       d. For the threaded end of a stud bolt which is screwed into the engine block.

5. Above what fraction do the unified thread sizes change to fractions?
       a. 1/8 in.   b. 1/4 in.   c. 1/2 in.   d. 3/4 in.

3-11:   CONVERSATION

Directions:

Lucy Garcia and Omar Harris are learning the names of some fasteners. Listen to their conversation and fill in the missing words. Rewind and practice your pronunciation.

Lucy: Omar, can you identify these _____?
    I'll point to them, and you _____ me their
    names. What's this with _____ threads?

Omar: That's a "hex-head bolt," or just "hex bolt."

Lucy: And, Omar, what does the word "_____" mean?

Omar: That's short for "hexagonal" which means "_____ angles," so it's a shape
    with _____ angles and six straight sides. Below the hex bolt is a hex _____;
    its interior _____ will fit on the _____.

Lucy: And what are those other three fasteners?

Omar: The one next to the hex bolt is an oval-head screw; the second _____ is a flat-
    head screw; the third one is a round- _____ screw.

Lucy: And notice those _____ in the heads; that's where the screwdriver blade fits.

The Parts of the Unified Thread Form

Directions: Test yourself on the names of the parts of the unified thread forms. (You may want to review before you take the test.) Look at the picture. Listen to the names on the tape. Write the letters of what you hear next to the numbers.

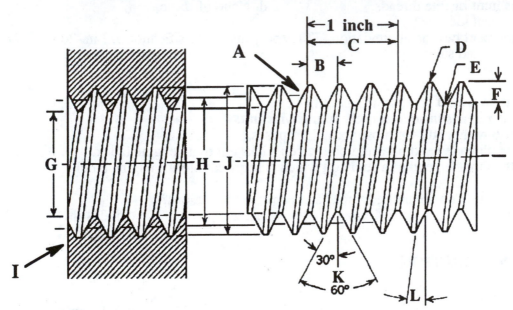

Audio Quiz:

1. _____   8. _____
2. _____   9. _____
3. _____   10. _____
4. _____   11. _____
5. _____   12. _____
6. _____   13. _____
7. _____   14. _____

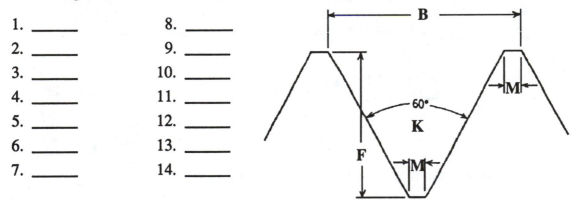

3-13: READING

Directions: There are many other things to learn about fasteners from practice in the shop and from reading your textbook. Be sure to read the topics below; put a check before each when you've read it. Drawing small pictures helps learn them.

Textbook pages:

___ p. 16: square heads and hex heads
___ p. 16-17: machine screws
___ p. 18: bolt markings for strength

___ p. 20-21: more about nuts
___ p. 23: washers
___ p. 24-25: pins, retaining rings, and keys

Unit 4: *SHOP DRAWINGS*

ASSIGNMENT: **Read, study and complete pages 21 to 40 of this book. Then read pages 26 to 34 in the textbook,** *Machine Tool Practices,* *SECTION A, Unit 3, Reading Drawings.*

OBJECTIVES for this unit:
You should be able to:
1. Identify different kinds of drawings used in the shop.
2. Identify different kinds of lines used in detail drawings.
3. Read and interpret two- and three-view orthographic projections, along with dimensions, tolerances, and standard abbreviations.

4-01: VOCABULARY LIST A
Directions: Study the vocabulary; write the missing words in the blanks.

1. to communicate — to exchange understanding by using words, pictures, or other means of expression. *Example:* Engineers need _____ their ideas for parts through drawings.

2. to interpret — to see a picture or an idea, and then show in what way you understand it. *Example:* A machinist looks at a drawing and is able _____ it by making a finished part.

3. to visualize — to see something in the mind. *Example:* Carlos was able _____ the finished part by looking at the drawing.

4. to intersect — to cross each other. *Example:* Plane A is able _____ plane B at the black line.

5. exploded view — a drawing showing the proper relationship between the parts of an assembly. *Example:* Bao saw an _____ of the washing machine assembly.

6. symbol — a sign or picture which means something beyond itself. *Example:* A plus or minus sign (±) is a _____ that stands for "add or subtract."

7. dimension — any measured length, such as the height, the width, or the depth, shown on a drawing. *Example:* It is very important to get the correct _____s on the parts you make.

8. perspective — a type of realistic drawing in which things become smaller as you look farther into a picture. *Example:* This picture uses _____ to show the railroad tracks.

(San Diego CCD)

9. maximum — the highest amount possible. *Example:* There are a _____ of six views for most solid objects.

27

4-02:   READING
Directions: *Read the following paragraphs.*

Communicating and Visualizing in the Shop

Machine shop students will soon be working for manufacturing companies. They need to make parts quickly and accurately. To do that, the worker needs information about a particular part. What shape is the part? How big is it? What's the part made of? The worker will get information like this by using *shop drawings*. These drawings are made by engineers and others who have designed the parts and decided how the parts are to be made. The drawing becomes an important way for engineers, machinists, and others in the shop to *communicate* about making the parts. There is a language of words and symbols that goes with these drawings. Learning this terminology is an important objective in this course.

When the machine shop student is able to read and interpret a shop drawing, he or she will be able to *visualize* a completed part.   That means, with the information given in the drawing, the worker will be able to see a completed part in his or her mind. The information from the drawing will guide the worker as he or she makes the part.

**Prints help
the machinist
to visualize**

There are two kinds of shop drawings used to communicate and visualize:

> 1. *an assembly drawing*: This kind of drawing shows a plan of <u>many</u> parts fitting together to make a completed machine; for example, an automobile engine.
> 2. *a detail drawing*: This kind of drawing shows all the information needed to make <u>one</u> individual part.

Detail drawings are used most often in the shop. Now look in the textbook; find the picture of a simple detail drawing; then find a picture of an assembly drawing.

4-03:   EXERCISE
Directions: *Read the questions. Circle the <u>best</u> answer.*

1. Shop drawings help the machinist to:
    a. visualize a finished part.   c. make the part quickly and accurately.
    b. communicate with engineers.   d. all of the above.

2. *To visualize* something means:
    a. to talk with others in the shop.   c. to read how to make the part.
    b. to see the part in your mind.   d. to make the completed part.

3. What kind of information does a shop drawing give the machinist?
    a. The dimensions of the part.   c. What the part looks like.
    b. The material used to make the part.   d. all of the above.

4-04: READING
Directions: Read the following paragraphs.

Different Kinds of Drawings

There are several kinds of drawings that can be used to show what a part looks like.

1. The Exploded Drawing

This kind of drawing shows the parts which make up an assembly. They are shown separately, as if the assembly had "exploded" into its individual parts. Shop workers making parts will not use this kind of drawing. Assemblers and repair persons will find this kind of drawing valuable.

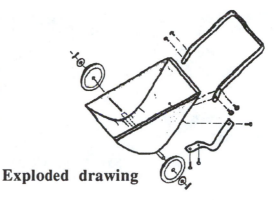

Exploded drawing

2. The Isometric Drawing

This drawing shows the length, width, and height of an object in one picture. It's called *isometric* because dimensions are always given in units of the same length, not as in a *perspective drawing* in which things farther away are drawn smaller. Isometric drawings are sometimes used to show how a completed part will look in three dimensions. The perspective drawing will not be used in the machine shop, but it helps us understand isometric drawings that do not shorten objects that are farther away.

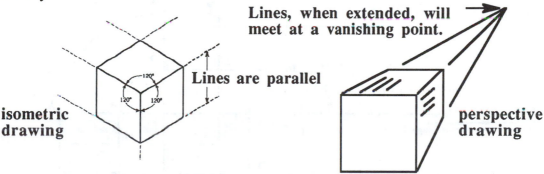

Lines, when extended, will meet at a vanishing point.

Lines are parallel

isometric drawing

perspective drawing

3. The Orthographic Projection

The orthographic projection is the drawing most often used in the shop. It shows the important features of a part. It does not have a three-dimensional look, but gives flat views from different places looking at the part. Two or three views are usually enough to show what is needed to make the part. There are a maximum of six possible views for any object: front, back, top, bottom, left side, and right side. The three usually used are the *front*, *top*, and *right side*.

Where does the name "orthographic projection" come from? Imagine a glass box with six sides. An object is in the middle of the box. Then one side of the object is projected out onto the nearest side of the box. In this way all six sides are projected, just like a movie image is projected onto a movie screen. The word *orthographic* means "drawn straightly."

ORTHOGRAPHIC PROJECTION

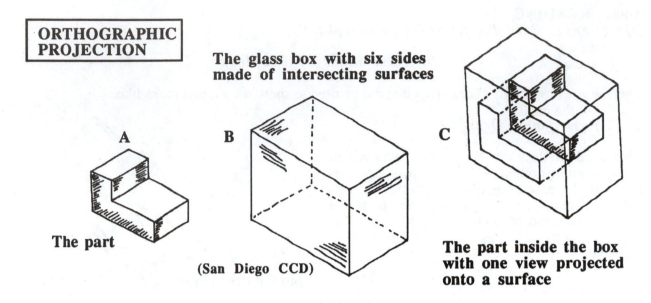

The glass box with six sides made of intersecting surfaces

A — The part

B (San Diego CCD)

C — **The part inside the box with one view projected onto a surface**

When the six sides are projected onto the sides of the box, the box is opened, and all six views can be seen. Three views are chosen to give a complete picture of the part.

In the picture below, notice how the points in one picture curve when they go into the next view; for example points A and B in the top view become points X and Y in the right side view.

SIX PROJECTED VIEWS

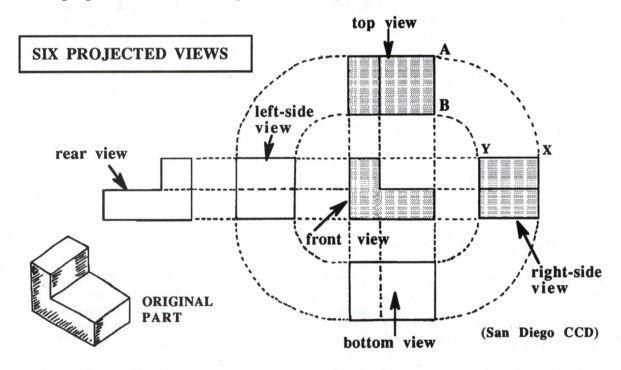

top view

left-side view

rear view

front view

right-side view

bottom view

ORIGINAL PART

(San Diego CCD)

There is a usual **ORDER** in which the three views are shown: <u>The front view is in the middle of the drawing, the top view is above the front view, and the right side view is on the right side of the front view.</u> This order should be followed in making and reading a drawing.

30

4-05: EXERCISE
Directions: Read the questions. Circle the __best__ answer.

1. The drawing at the right is an example of:
   - a. exploded drawing.
   - b. isometric drawing.
   - c. perspective drawing.
   - d. orthographic projection

2. What is the maximum number of views that could be projected onto the sides of a glass box?
   - a. two
   - b. three
   - c. five
   - d. six

3. What is the usual number of views used to show a part in an orthographic projection?
   - a. two
   - b. three
   - c. five
   - d. six

4. What kind of drawing has a vanishing point?
   - a. an exploded drawing.
   - b. an isometric drawing.
   - c. a perspective drawing.
   - d. an orthographic projection.

5. What kind of drawing is usually used to show the main features for making a part?
   - a. an exploded drawing.
   - b. an isometric drawing.
   - c. a perspective drawing.
   - d. an orthographic projection.

6. What views are usually shown in an orthographic projection drawing?
   - a. front, top, right side.
   - b. front, bottom, left side.
   - c. top, bottom, right side.
   - d. top, front, left side.

7. A corner of a rectangular part in an isometric drawing is shown by three angles that are each:
   - a. 120°
   - b. 90°
   - c. 60°
   - d. 30°

4-06: EXERCISE:

Directions:
Carlos made a tape recording of a lecture in class. He had also taken notes, but hadn't been able to write down all the key words. Listen to his recording of the teacher and help him fill in his notes.

Several kinds of drawings (d's): _____ d's, _____ d's and

_____ d's. One most often used in the shop = _____

projection. Shows 2 or 3 _____ . Enough to visualize part. 3 usual _____ =

front, _____, & _____side. Iso_____ d's look more _____,

because they show more than one _____ of part at one time. Leading corner is made with 3

angles of _____°. Difference between these and per_____ d's is that in a

per_____ d, farther into picture you go, the _____ things look, but in an

iso_____ d all _____ are the same. For next time study pp. _____

in TB. Fri. quiz on _____ of lines.   (nomenclature = technical names).

31

Directions: *Study the vocabulary; write the missing words in the blanks.*

1. visible   able to be seen. *Example:* Not all edges are _____ in a drawing of a part.

2. dash   a short thin line; in shop drawings dashes line up in a row. *Example:* To show hidden edges, use a row of _____ es .

━━ ━━ ━━ ━━ ━━ ━━ ━━ ━━
   a row of dashes

3. to alternate   to take turns; to do something every other time. *Example:* To draw a center line, it is necessary _____ dashes with long lines.

━━ ━━━━━━━━━ ━━ ━━━━━━ ━━ ━━━━━━ ━━

   Dashes are alternated with long lines to make a center line.

4. fillet   a curved surface joining two intersecting surfaces. *Example:* This _____ has a 1/2 inch radius.

fillet

5. to extend   to make something longer. *Example:* Bill drew a line _____ from point **A** past point **B** and half way to point **C**.

   A   **B**   **C**

6. segments   the parts into which a line can be divided. *Example:* A dimension line has two _____ with the length number in between them.

◄─────────── 2.750 ───────────►
   dimension line with two segments

7. detail   an important part of a part. *Example:* This _____ of the drawing has dimensions of 1.500 and 2.000.

8. wavy   in the shape of a wave; curved, not straight. *Example:* Here are two lines: one is straight, the other is _____.

━━━━━━━━━   ∿∿∿∿∿
   straight line   **wavy line**

9. chamfer   a cut made along the edge of a part to remove the sharpness of the edge. *Example:* A _____ is usually cut at a 45° angle.

45°

chamfer

$\frac{1}{8}$

10. uniform   having the same shape or appearance along the length of an object. *Example:* This pencil has a _____ shape along its length.

4-08: READING
Directions: Read the following paragraphs.

Kinds of Lines Used in Drawings

A necessary skill in reading and interpreting shop drawings will be to know what lines you are seeing in a particular drawing. There are several kinds of lines; each has a different purpose.

On the right is an isometric drawing of a part. To make this part, the machinist will look at a three-view orthographic projection of front, top, and right side. Notice how the surfaces shown in the isometric drawing transfer to the three views.

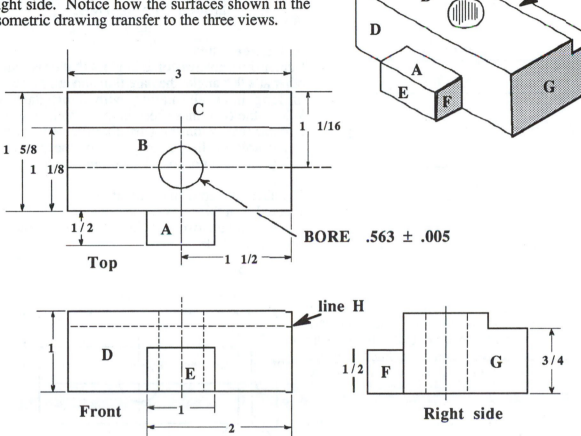

BORE .563 ± .005

Directions: After reading about each line, copy its name on a piece of paper.

1. Visible lines or object lines
Visible lines show the visible edges of the part. A visible or object line is a thick, unbroken line. In the drawing above, the object lines are the darker lines that show the basic shape of the part. They are visible in all three drawings. At right, the right-side view is shown without the hidden and other lines. Notice how surfaces **F** and **G** come over from the isometric drawing. In the top view, the edge of the hole is also shown by an object line.

Line H will not be visible in the front view

**Right side
Object lines**

33

2. Hidden lines

These lines are used to draw those features of the part which are not visible when you look at a view. A hidden line is a medium thick line made up of short dashes.

In the front view shown at the right, the line *P* (dashes) is a hidden line showing an edge which can only be seen from the sides or the rear. The lines marked *Q* show the hidden edges of the hole which can only be seen in the top view.

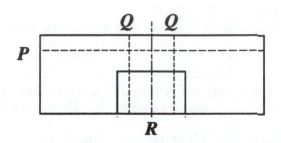

Front View: Lines *P* and *Q* show features that are hidden in this view.

Top View

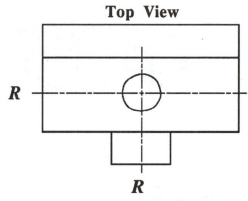

Lines marked *R* are center lines

3. Center lines

Center lines always come in pairs that cross each other at a 90° angle; they are thin and are made up of long lines alternating with short dashes. Where the two lines cross is the center of some circular feature on the part, like the center of a round hole or the center point for measuring a radius on a fillet or rounded corner or edge.

The line marked *R* in the front view and the lines marked *R* in this top view are center lines. The place where the two lines intersect is the center point of the hole.

4. Extension lines

These lines are thin and continuous (unbroken); they show the continuation of some important feature. They are often extended out from the object lines, but do not touch those lines; they are extended so that dimensions can be shown more easily.

The lines marked *S* are examples of extension lines.

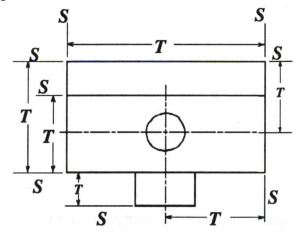

5. Dimension lines

Dimension lines are thin with an arrowhead at each end. They are most often broken in the middle so that the dimension length can be written between the two line segments. The dimension lines are drawn parallel to the drawing line for which you wish to give a dimension. In the picture above, the break in each dimension line is marked with the letter **T**. This is where the dimension numbers are written in the real drawings.

If the space between two extension lines is too narrow, the two arrows can be drawn pointing *in* at the extension lines, as shown in the drawing at the right.

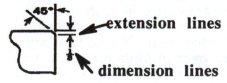

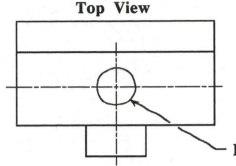

Top View

6. Leaders

A leader is a thin line with an arrowhead at one end used to point to an important detail in a drawing. The other end leads to notes that give more information about the area.

In this example from the top view, a note gives the size of the hole to be bored.

BORE .563 ± .005

7. Break lines

A short break line is a thick, wavy line that has two uses:

    a. It can show the details of some larger part

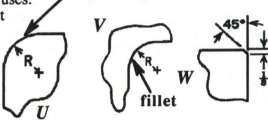

radius

In picture *U*, a corner from a larger part is shown; the break line is the wavy line near the letter; the picture also shows two center lines from which the radius to the rounded corner of the part is drawn.

In picture *V*, the break line is again a wavy line near the letter; an inside radius is shown on the other side; again the intersection shows the point from which to measure the radius.

In picture *W*, the short break line is the wavy line near the letter; a chamfer is shown in the upper right hand corner. The chamfer is marked with a 45° angle mark and its height is shown with two extension lines. This height is 1/8 inch; the two lines with arrowheads did not fit between the extension lines, so they are placed pointing *in* toward those lines.

    b. Short break lines can shorten a part that is uniform throughout its length.

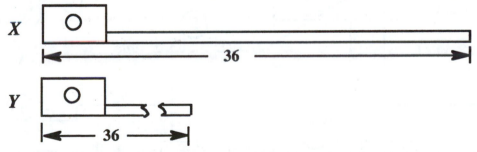

Both *X* and *Y* are accurate pictures of the same view. Picture *Y*, however, uses a short break line to show how the long uniform piece has been shortened. Note that the dimensions given for both parts are the same.

A round stock break uses two thick lines that are each shaped like an "S." It is used in a drawing to shorten a piece of round stock that is uniform through out its length.

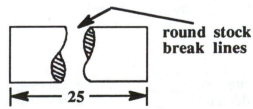

round stock break lines

8. Section lines

When a cut is made through a part, the inside may look different than the outside. Section lines are used to show what that inside looks like. Section lines are thin lines usually drawn at a 45° angle to the horizontal. Such a view can be called a "cross section," or a "sectional view."

In the example on the right, a group of circles in the upper drawing is made more clear by showing the cross section AA.

9. Cutting Plane lines

In the upper drawing the line AA is a cutting plane line; it shows where the cut was made down through the part. The arrowheads point in the direction toward which the viewer is looking when seeing the cross section.

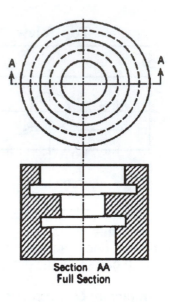

Section AA
Full Section

4-09: EXERCISE

Directions: Match the letter of the line with the number of the name.

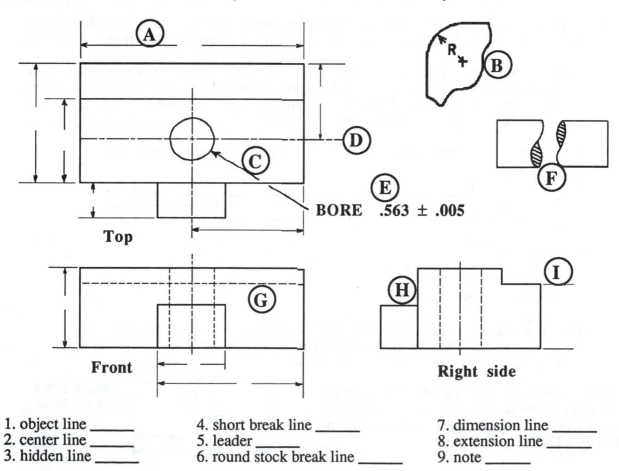

1. object line _____
2. center line _____
3. hidden line _____

4. short break line _____
5. leader _____
6. round stock break line _____

7. dimension line _____
8. extension line _____
9. note _____

36

Directions:
Practice asking questions to get the information you need from your instructor and from other students.

One way to ask questions is to use question words like these:

Who?	**What?**	**Which?**	**When?**	**Where?**
Why?	**How?**	**How many?**		**What kind?**

For this exercise, start with an answer that is in the form of a statement; then listen to a question that goes with that answer. <u>*Write what you hear*</u> .

Examples:
Answer: That's a hidden line.
Question: <u>*What's the name of that line with dashes?*</u>

Answer: We're studying nine kinds of lines.
Question: <u>*How many lines are you studying?*</u>

1. Answer: Because you need <u>two</u> intersecting lines to mark the center of the hole.
   Question: _____

2. Answer: When you want to show some detail of a larger part.
   Question: _____

3. Answer: Dimension lines are drawn between two extension lines.
   Question: _____

4. Answer: Engineers, machinists and other shop workers use them.
   Question: _____

5. Answer: Use a cutting plane line; then show a sectional view using section lines.
   Question: _____

6. Answer: Use them to connect a note with some part of the drawing.
   Question: _____

7. Answer: Object lines are used to show that. They're also called visible lines.
   Question: _____

8. Answer: Extension lines and a dimension line with the length number.
   Question: _____

9. Answer: Each broken end is shaped like an "S."
   Question: _____

4-11: VOCABULARY LIST C
Directions: _Study the vocabulary; write the missing words in the blanks._

1. exact   very accurate without varying from a given number. _Example_: The _____ dimension is the perfect dimension.

2. nominal   agreeing in value with a given number. _Example_: The _____ size of this part is 6.500 inches; the real size might actually be a little smaller (for example, 6.499) or a little larger (for example, 6.501).

3. acceptable   meeting a certain value that was set ahead of time. _Example_: For a dimension of 6.500 ± .003, the actual dimensions of 6.501 and 6.498 are _____ sizes.

4. bilateral   having two sides. _Example_: A dimension of 3.455 ± .005 has a _____ tolerance, because the tolerance moves in _two_ directions, an upper limit of 3.460 and a lower limit of 3.450.

5. scrap   a part that has a dimension made outside of tolerance; it's too large or too small. _Example_: Efren is very careful at work; he never makes parts that are _____.

range =
.745 to .755

.750 ±.005

.759 (It's too long, but this part could be reworked to remove more material.)

.743 (It's too short, and you can't put the material back. It's scrap! Recycle it!

6. to represent   to be or act in place of something or someone else. _Example_: The views in the drawing are meant _____ real parts which the machinist makes.

7. smooth   having an even, level surface with no roughness. _Example_: The surface of the part was very _____ after it went through the grinder.

8. microinch   one millionth of an inch; the _micro-_ part of the word means "a millionth of." _Example_: Smoothness of a surface is given by the average height of the bumps on a surface measured in _____ es .

9. bearing   a supporting part that carries the weight of another part. _Example_: The surface inside a _____ needs to be very smooth; that will lessen the amount of friction with the part rotating inside.

10. unilateral   having one side or measurement in only one direction. _Example_: The diameter of this fitting cannot be larger than 1.000 inch, but it could be .999 or .998; that dimension has a _____ tolerance.

4-12: READING
Directions: Read the following paragraphs.

Other Information about Drawings

1. Tolerance

To make parts, the machinist uses the inch and fractions of an inch, written as bar fractions or decimals. A dimension of six and one half inches could be written *6 1/2* or *6.500*.

For a part which has a dimension of 6.500 given, the 6.500 is the *nominal size*--the size called for by the print. Acceptable parts will be those which are within a range of lengths around the nominal size. This lets the machinist know how much room there is to vary slightly from the nominal size. This acceptable variation from the nominal size is called *tolerance*. Tolerance is shown by a ± (plus or minus) sign written after the nominal size, as in the top view of the orthographic projection shown earlier: "BORE .563 ± .005."

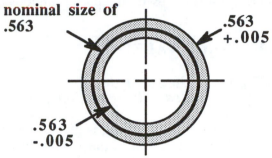

In this example, when .005 is added to .563, the result is .568; when .005 is subtracted from .563, the result is .558. This means that any dimension between .568 (the upper limit) and .558 (the lower limit) is acceptable. Parts outside these limits are called *scrap*.

Bore .563 ± .005

This kind of tolerance is called *bilateral*, because the range of acceptable parts goes in *two directions*, above and below the nominal size. Sometimes the tolerance only goes in *one direction,* as in the example 1.000 +.000, -.002. On one side of 1.000 there is no room for the dimension to be larger; on the other side, the dimension could be .999 or .998; this kind of tolerance is *unilateral* (with one side).

acceptable sizes are:
.558, .559, .560, .561, .562, .563, .564, .565, .566, .567, and .568.

2. Scale

The relation between size in the drawing and size of the actual part is called *scale*. Scale is shown on drawings in a special box called "scale."

Views in a drawing can be the same size as the part they represent; in that case, the scale is called FULL or 1 = 1 (read "one to one"). At other times the views are smaller than the actual part; for example, a drawing view may be 1/2 the actual size; the scale in that situation is 1/2 = 1, which means that 1/2 inch on the paper represents 1 inch on the actual part.

Sometimes the view is larger than the actual part, especially when representing a very small part; for example, two inches on the drawing are used to show one inch on the real part; the scale would be 2 = 1. In all situations, *the measurement on the drawing is given first, and the measurement on the part is given second.*

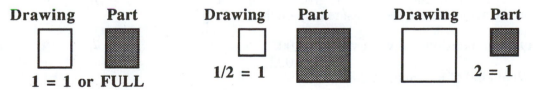

3. Finish marks

The smoothness on the surface of a part is called *finish*. Shop drawings sometimes include directions on how smooth the surface of a part is to be made. Smoothness is measured by how rough the surface is by measuring the average height of the bumps that vary from a flat surface; these tiny heights are measured in *microinches*. One microinch equals one millionth (.000001) inch.

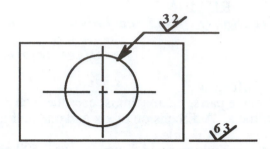

Numbers 4, 8, 16, 32, 63, 125, 250 are measurements of surface smoothness in microinches. The smaller the number, the smoother the surface. The number is shown on the drawing along with a finish symbol, which looks like a *V*. The picture shows a finish of 63 on the exterior surface and a finish of 32 on the inside of the hole; the hole is made less rough, because it will be a bearing for some revolving part; a smoother surface will cut down on friction.

4-13: "SHOP TALK" QUESTIONS AND ANSWERS

Directions:
Read the questions, listen to the tape, and write the answers.

1. A dimension in a drawing is given as 3.750 ± .003. Some parts were made and measured. Write down the letters of those parts that are within tolerance:

    _____

2. Compute the limits for a dimension of 2.753 with a tolerance of _____
    Upper limit (maximum limit) = _____
    Lower limit (minimum limit) = _____

3. In a drawing, a sliding surface has a finish of 32, a bearing has a finish of _____, and the surface of the front has a finish of _____. Which one has the smoothest finish? _____

Directions:
Read the questions, listen to the tape, and circle the __best__ answer.

4. Q: What are the parts that are *not* acceptable called? A B C

5. Q: Length of actual part = 6.500 inches. How long would the dimension be on a drawing be with a scale of 1/2 = 1? A B C

6. Q: What is the size of the dimension given on the drawing called? A B C

7. Q: What do you call a tolerance which is added and subtracted from the given dimension? A B C

8. Q: For showing scale, which is given first, the measurement on the actual part or the measurement on the drawing? A B C

9. Q: A dimension is given as 2.533 + .000
                                       - .002
    What kind of tolerance is this? A B C

Unit 5: *HAND TOOLS*

ASSIGNMENT: **Read, study and complete pages 41 to 50 of this book. Then read pages 43 to 82 in the textbook,** *Machine Tool Practices,* *SECTION B, Hand Tools.*

OBJECTIVES for this unit:
You should be able to:
1. Identify and say correctly the names of some common hand tools.
2. Identify the purpose and correct use of these tools.
3. Pronounce correctly the nomenclature of the tools and use the terms in model conversations.

5-01: VOCABULARY LIST A
Directions: Study the vocabulary.   Write the missing words in the blank spaces.

1. invention   any mechanical product that has been thought out and made for some particular purpose. *Example:*   Travel became easier because of an _____ called the automobile.

2. development   the process of taking an invention and changing it to make it better. *Example:* An important _____ in today's automobiles is the addition of computers to control the flow of gasoline and air.

3. device   a mechanical invention made for some purpose. *Example:* The C-clamp is a useful _____ for fastening work to a drilling table.

   C-clamp
   (SDCCD)

4. to tighten   to put something together closely and securely. *Example:* Omar, be sure _____ the bolts on that machine.

5. to loosen   to unfasten, to remove tightness. *Example:*   Kyko had _____ the jaws to take the workpiece out of the vise.

6. to grip   the act of holding onto something tightly. *Example:*   It is necessary _____ the handrail tightly if you stand up on the bus.

7. joint   a part where two things are fastened together. *Example:* The handles of a pair of pliers have a _____ that holds them together.

   joint
   pliers

8. diagonal   **A**   a straight line running from one opposite corner to the other. *Example:* In this picture you can see a _____ line running from A to B.   **B**

9. portable   able to be carried easily from one place to another. *Example:* Pliers, screwdrivers, and files are examples of _____ tools.

10. to force fit   to push a part into a hole which is slightly smaller than the part going in. *Example:* Jerry _____ _____ the bushing into the hole.

41

5-02: READING
Directions: Read the following paragraphs.

SOME COMMON HAND TOOLS

The invention and development of better tools is the reason why modern life has become more convenient. Today we can fly from coast to coast in hours and drive from one part of the city to another in a few minutes. Machine tools have improved transportation and every material aspect of our lives.

You will soon study the large machines like lathes and milling machines; first, however, be sure you know the names and correct use of some common **hand tools** which are used by modern machinists. There will be times when you need them; learn now how to correctly pronounce their names, so you can ask for them at the tool room. These tools are called "hand tools" because they are small and portable. Hand tools can be classified by their uses: holding, fastening, and cutting.

A. Hand Tools for Holding

The **bench vise** is a common holding device in the machine shop. It is attached to the machinist's workbench; it has a pair of steel jaws for holding a workpiece; the jaws are tightened around the workpiece by turning a handle.

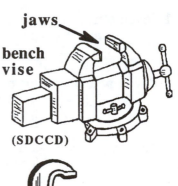

jaws

bench vise

(SDCCD)

The machinist also has need sometimes for clamps; a common kind is the **C-clamp** which will hold a workpiece on a drill press table or hold two larger pieces together. The clamps have handles to turn to tighten or loosen the grip.

C-clamp
(SDCCD)

Pliers are another useful holding tool. Very commonly used is the **slip-joint pliers** with two positions for its jaws, depending on the size of the what you are holding. **Needlenose pliers** will allow you to grip a part in difficult-to-reach places. **Diagonal cutters** look like pliers with their two handles, but this tool is not for holding; it's for cutting wire or thin pieces of soft metal.

slip-joint pliers   needlenose pliers   diagonal cutters

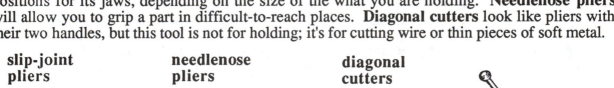

B. Hand Tools for Fastening

The **arbor press** is a tool usually mounted on a work bench used to "force fit" one part into another; it allows the machinist to concentrate pressure on parts like a bushing (a bearing) which must be forced into a pre-drilled hole or to press a shaft into a hub (the center of a wheel). Results will be best if the holes have chamfers, if lubricants (like oil) are used, and if the machinist can, with practice, "get a feel" for the correct amount of pressure to apply.

arbor press

5-03: EXERCISE
Directions: Read the questions. Circle the best answer.

1. What tool is used to hold a workpiece on a drill press table?
   - a. slip-joint pliers
   - b. C-clamp
   - c. bench vise
   - d. an arbor press

2. What tool is used for grasping parts in difficult-to-reach places?
   - a. bench vise
   - b. arbor press
   - c. needlenose pliers
   - d. diagonal cutters

3. Which of the following will help in the use of the arbor press?
   - a. chamfering the hole into which you wish to press something
   - c. use of lubricants on the hole and the piece
   - c. getting a feel for the correct amount of pressure to apply
   - d. all of the above.

4. Which of these tools would you use to cut wire or sheets of soft metal?

   a.    b.    c.    d.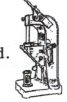

5. Which tool has two positions for holding objects of different sizes?
   - a. slip-joint pliers
   - b. C-clamp
   - c. bench vise
   - d. needlenose pliers

6. Which tool has two steel jaws for holding objects, a handle for tightening the jaws, and is attached to the bench?
   - a. slip-joint pliers
   - b. C-clamp
   - c. bench vise
   - d. arbor press

5-04: "SHOP TALK"   *ASKING FOR HELP and GETTING DIRECTIONS*

 Directions: Listen to the students asking for help; then listen to three possible directions from the teacher. Circle the best teacher response. Practice the pronunciation of the sentences.

Student Request: *Teacher Response:*

1. I need to force fit this bushing into this hole.   A   B   C
Please show me what tool to use.

2. Can you please tell me the steps for putting in a bushing?   A   B   C

3. I want to make sure the workpiece won't move when   A   B   C
I start drilling. Can you show me how to attach it to
the drill press table?

4. I want to fit this shaft into this hub. Please show me how.   A   B   C

43

5-05: VOCABULARY LIST B
Directions: *Study the vocabulary. Write the missing words in the blank spaces.*

1. adjustable   able to be changed so it will fit. *Example:* The jaws of this wrench are _____ so they will fit on many sizes of bolts.

2. socket   a hollow part into which something fits. *Example:* One end of the _____ fits over a bolt or a nut; the other end of the _____ receives a drive which turns it.

socket

drive   (SDCCD)

3. drive   a device that communicates motion to another part. *Example:* When the wrench is turned the _____ turns the nut.

4. set   a collection of tools which are like each other, but are of different sizes. *Example:* Pham's wife bought him a _____ of sockets and a socket wrench for his birthday.

socket wrench   **socket set**

(San Diego CCD)

5. to mount   to put, or fasten, in the proper place. *Example:* The saw is not sharp; I need _____ a new blade in the frame.

frame

hacksaw   **blade**

6. burr   a rough edge left on a metal part by cutting or drilling. *Example:* You can use a file to remove a _____ from a workpiece; this process is called "deburring."

7. flute   the groove in a drill, tap, reamer, or milling cutter. *Example:* A drill can have one or more _____s. (See page 355 of the textbook.)

flute

shank   **drill bit**

8. shank   the part of a tool which is held in a toolholder or in the hand. *Example:* The _____ of the drill is not as long as the body with its flutes.

9. stroke   a movement forward of the hand holding a saw or a file. *Example:* A file or saw cuts during the forward _____.

10. curve   a line having no straight part or angled part. *Example:* This part has a large interior _____ which must be filed.

interior curves

44

5-06: READING
Directions: Read the following paragraphs.

MORE COMMON HAND TOOLS

B. Hand Tools for Fastening (continued)

Wrenches of different kinds are very useful for fastening nuts and bolts and turning other threaded pieces. An **open-end wrench** or a **box-end wrench** is useful for turning hex or square head bolts and nuts. An **adjustable wrench** (also called **crescent wrench**) has a screw with which the jaws can be opened to fit around fasteners of different sizes, but the fit is not very secure sometimes. **Allen wrenches** (or **socket head wrenches**) have hex-shaped ends which fit into hex-shaped slots in the heads of certain screws.

(Ssn Diego Community College District)

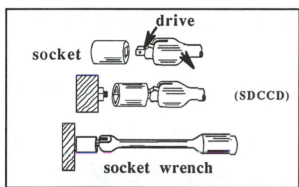

A **socket wrench** is like the box-end wrench, because it goes completely around the nut or bolt; the gripping pieces come in a **socket set** of different sizes, each with a socket at one end and a drive end at the other; the socket wrench has a square-shaped drive which fits into the drive end. The socket sizes can be from either the U.S. Customary system (using inches) or the International Metric system (using millimeters).

Screwdrivers are used to tighten or loosen screws. They are of two types: the **standard** and the **Phillips**; the tips of the blades fit into the slots of standard or Phillips screws. The standard screw has a slot like a minus sign; the Phillips screw has slots like a plus sign or a star.

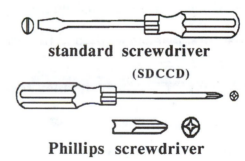

5-07: EXERCISE
Directions: Circle T (for "true") or F (for "false") for each of the following statements.

T	F	1. Sockets sizes can be either metric or the U.S. Customary.
T	F	2. The drive end of a socket surrounds the nut or bolt, like a box-end wrench.
T	F	3. The names socket head wrench and Allen wrench describe the same thing.
T	F	4. Open-end wrenches go easily around hex nuts and hex bolts.
T	F	5. The slot of a standard screw looks like a plus, the Phillips like a minus.
T	F	6. The adjustable or crescent wrench is the best tool for removing hex nuts.

AND MORE HAND TOOLS

C. Hand Tools for Cutting and Filing

The **hacksaw** is used for cutting metal which is secured firmly in a bench vise. The saw has a blade mounted in a frame which connects to a pistol-grip handle.   The teeth of the blade point away from the handle and cut only on the forward stroke; coarse-tooth blades are usually used to cut soft materials and fine-tooth blades are used for harder materials.

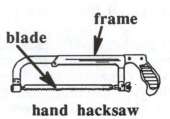

hand hacksaw

Files are hand tools used for purposes like these: to remove burrs and sharp edges from workpieces, to shape a workpiece so it will fit well, and to put a finish on a part. The main parts of a file are shown here; the **tang** of the file fits into a **handle**; if the file slips while in use without a handle, the tang can injure the hand or wrist. Other main parts of a file are **length**, **point**, **face**, **edge**, and **heel**.

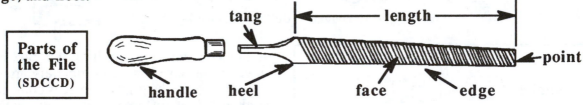

Parts of the File (SDCCD)

There are four different **cuts** of files: **single**, **double**, **curved tooth**, and **rasp**. Files also vary in their coarseness. (See the pictures on pages 55-57 of the textbook.)

Some commonly used types of files are these:

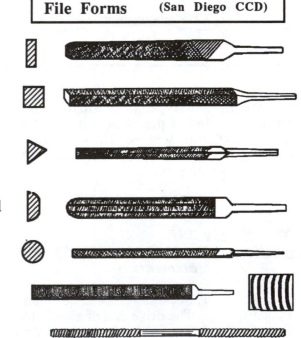

File Forms   (San Diego CCD)

flat files are usually double-cut and can be used to file on lathe or filing a finish on a workpiece;

square files, that can be used to file keyseats, slots, and holes;

three-square files (or **triangular files**) are used to file internal angles of 60° to 90° in a workpiece or make other sharp corners;

half-round files can be used to file large internal curves or shape the radius on a rounded corner;

round files are used to enlarge holes;

curved tooth files will remove material very rapidly;

thread files are used to reshape damaged threads; often a lathe is used to turn the threaded piece.

Files only cut on the forward stroke. Look carefully at Figure B-98 in the textbook; it shows the correct way to hold a file.

EXERCISE
Directions: *Write the letter of the statement next to the number of the word the statement matches.*

____	1. half-round file	a. has a frame and a blade for cutting metal
____	2. curved tooth file	b. the pointed end of a file which fits into a handle
____	3. thread file	c. usually used to cut soft materials
____	4. tang	d. used to file large internal curves or to shape a radius
____	5. round file	e. can be used to file on a lathe or finish a workpiece
____	6. square file	f. used to reshape damaged threads
____	7. triangular file	g. used to make keyseats, slots, and holes
____	8. hacksaw	h. can file internal angles of 60° to 90°
____	9. flat file	i. used to enlarge holes
____	10. coarse tooth blade	j. will remove material very rapidly

5-10: READING
Directions: *Read the following paragraphs.*

ABOUT REAMERS, TAPS, AND DIES

D. Reaming Holes and Cutting Threads

Another hand tool is the **hand reamer** used to finish a hole made by a drill press; a drilled hole is sometimes rough but can be given a smoother finish and more accurate dimensions with a reamer. Reamers can be **straight** or **tapered** for reaming straight or tapered holes. At one end reamers have straight or helical **flutes** with sharp cutting edges; this fluted end of the reamer is chamfered a little to help the reamer start its cut. At the other end of the reamer is a **square** which fits into a **tap wrench** or a **T-handle**; between is the **shank** and the **neck**.

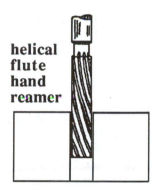

helical
flute
hand
reamer

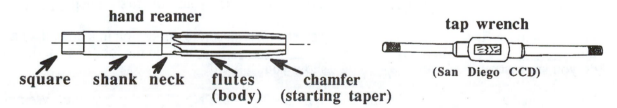

hand reamer

square   shank neck   flutes   chamfer
                        (body)   (starting taper)

tap wrench

(San Diego CCD)

Taps are used to cut internal threads on the inside of holes that are already drilled. Sometimes the hole goes all the way through the part. Other times the hole only goes part way through the part and is called a "blind hole." Taps can be of various sizes and usually come in **a set of three**. The first tap begins cutting the thread, the second cuts the thread deeper, and the third tap completes the threads.

(San Diego CCD)

47

5-11: READING
Directions: Read the following paragraphs.

D. Reaming Holes and Cutting Threads (Continued)

Dies are used to cut external threads on round materials, but should not be used on hardened or hard metal which will dull the cutting edges of the die. There are several types of dies:

Adjustable split dies use a small adjusting screw to force the sides of the die apart and make it slightly larger. These can also be put inside a **diestock** which has adjusting screws which will widen or narrow the opening of the split die.

diestock
(SDCCD)

two-piece dies (SDCCD)

Two-piece dies with two halves are held in a **collet** which is held in a diestock; it is important to make sure that the two halves you use are of the same size.

Rethreading dies (hexagonal or square) that are used to recut the threads on damaged or rusty bolts.

rethreading die
(SDCCD)

5-12: "SHOP TALK"   *ASKING FOR HELP and GETTING DIRECTIONS*

 Directions: Listen to the student asking for help; then listen to the teacher giving directions. Fill in the blanks. Practice the pronunciation of the sentences.

Student Request:

1. I want to recut the _____ on these bolts. They're very rusty.

2. And how do I put _____ threads on a round rod?

3. I want to thread the inside of this _____ hole. How do I start?

4. I need to have a fine _____ in this _____ I've drilled. What do you suggest?

5. I need to make a _____ in this gear.

6. I'm going to do some _____. Do you have any suggestions for me?

Teacher Directions:

Use a hexagonal _____ die. Make sure you have the right size, and put it in a _____.

Get a two-piece die and put the two halves around the rod. Put them in a _____. Then put the collet in a _____.

You'll use a set of _____ and a tap wrench for a job like that. Drill your hole first. Then _____ it.

To get a more accurate dimension in a hole, the hand _____ will do the job for you. Let me show you.

Why don't you use a _____ file. We have all the common _____ in the toolroom.

Why, yes, I do. Make sure you put the _____ in a handle. Use the correct file for the job. Remember that the cutting action is during the _____ stroke. And don't _____ the file over the work on the return stroke.

48

Hand Tools

Directions: *Study these pictures of some common hand tools, as you listen to the pronunciation of the name of each tool. Pronounce each name. Listen again, and write the names on another sheet of paper.*

1. bench vise
2. C-clamp
3. slip-joint pliers
4. diagonal cutters
5. arbor press
6. needlenose pliers
7. socket wrench
8. socket set
9. hacksaw
10. standard screwdriver
11. Phillips screwdriver
12. open-end wrench
13. box-end wrench
14. combination wrench
15. adjustable wrench
16. length
17. handle
18. tang
19. heel
20. face
21. edge
22. point
23. Allen wrenches
24. flat file
25. round file
26. half-round file
27. triangular file
28. thread file
29. square file
30. curved tooth file

31. hand reamer
32. tap wrench
33. rethreading die
34. two-piece dies
35. set of taps
36. diestock

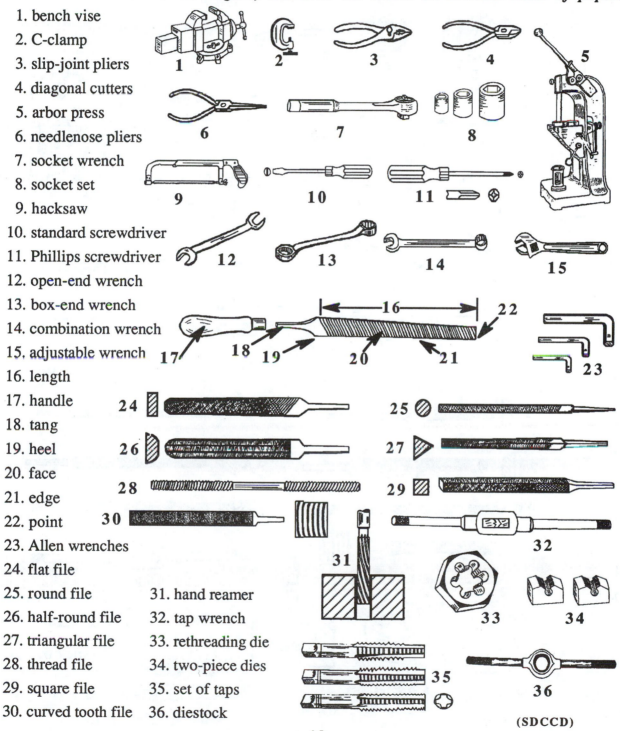

(SDCCD)

49

Hand Tools

<u>Directions</u>: *Test yourself on the names of these common hand tools. (You may want to review before your take the test.) Look at the pictures. Listen to the names on the tape. Write the letters of what you hear next to the numbers.*

1. ____
2. ____
3. ____
4. ____
5. ____
6. ____
7. ____
8. ____
9. ____
10. ____
11. ____
12. ____
13. ____
14. ____
15. ____
16. ____
17. ____
18. ____
19. ____
20. ____
21. ____
22. ____
23. ____
24. ____
25. ____   31. ____
26. ____   32. ____
27. ____   33. ____
28. ____   34. ____
29. ____   35. ____
30. ____   36. ____

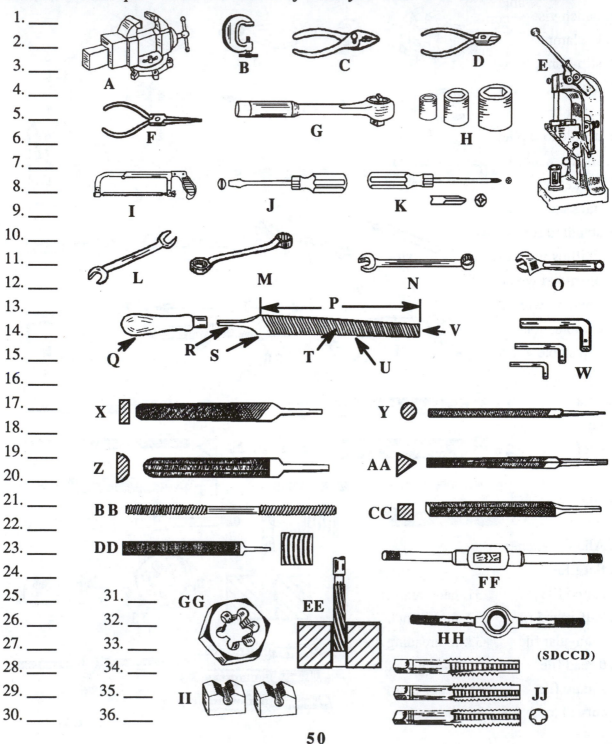

50

Unit 6: *THE LANGUAGE OF MATHEMATICS*

ASSIGNMENT: **Read, study and complete pages 51 to 68 of this book. Then look in the textbook at pages like: 101, 258-259, and 444-446.**

OBJECTIVES for this unit:
You should be able to:
1. Say and write correctly terminology from basic mathematics, geometry, and trigonometry.
2. Say and write correctly some abbreviations and formulas used in the machine shop.
3. Identify the trig functions on the scientific calculator and compare them with the trig tables.

6-01: VOCABULARY LIST   A
Directions: Study the vocabulary. Write the missing words in the blank spaces.

1. to compute

   to arrive at a value, number, or amount by working a math problem. *Example:* Are you able _____ the answer to this problem: **7 + 12 + 4** ?

2. formula

   a set of letters and numbers which tell about a mathematical relationship. *Example:* The _____ for computing the distance around the edge of a rectangle is P (perimeter) = 2L + 2W.

   > **Formulas:**
   > 1. RPM = $\dfrac{4CS}{D}$
   > 2. $C = \pi\, d$

3. section

   a part or division of a book, or a chapter, or a lesson. *Example:* Read the first two _____s ; then write the answers to these questions.

4. previous

   coming before in time or order. *Example:* I will repeat my _____ statement: There will be no school next Friday.

5. frequent

   happening often, repeatedly, at short intervals. *Example:* Mohammed is a _____ visitor to his old neighborhood in Rockville.

6. abbreviation

   a shortened form of a word or phrase. *Example:* "OD" is the correct _____ for the words "outside diameter."

7. quantity

   the measured amount of something, expressed in numbers. *Example:* Here are 150 new parts; that's a large _____.

8. mathematical operations

   any activity like adding, subtracting, multiplying and dividing which can be applied to quantities. *Example:* You can apply _____ to whole numbers, to fractions, and to decimals.

9. equivalent

   equal in quantity, value, or meaning. *Example:* The decimal _____ for the fraction 1/2 is .500.

10. lowest terms

    the results when a bar fraction is reduced until it can be reduced no further. *Example:* Jim reduced 4/8 to 2/4, but he didn't reach _____ until he further reduced the 2/4 to 1/2.

51

6-02: READING
Directions: *Read the following paragraphs.*

TALKING ABOUT MATHEMATICS

1. Do you know the language of math?

Computing math problems seems like a different job than speaking English; however there will be many times in your machining career that you will need to speak the language of math. Imagine that you are talking with a supervisor about some important dimensions on a part:

"How long is that dimension?"

"Is this tolerance correct? It seems too large to me."

"I need to compute the speed setting for this mill. What's the formula for circumference?"

Statements, questions, and problems from everyday work often involve English words used to talk about mathematics. In this section your goal is to make sure you understand some basic mathematics and have the English to talk about it.

2. Reading numbers

Some numbers sound very similar to each other, especially pairs like "fifty" and "fifteen." This is even more difficult for Asian and Spanish-speaking people who often leave the final sounds off of English words: If the "n" gets dropped from "fifteen," the word will sound just like "fifty." In this section, practice pronouncing numbers slowly and clearly with all the sounds that are there. You want to be understood when you speak and you want to understand your co-workers.

> • *Be careful to pronounce all the sounds in your numbers.*
>
> • *15 can sound like 50, if you don't pronounce the final "n" clearly.*

6-03: PRONOUNCING NUMBERS

Directions:

• *Items a to l: Look at each number below. Listen to each number on the tape. Pronounce each number slowly and clearly before the tape goes on to the next number. Make sure you're saying the final sound on each word.*

• *Items m to t: Listen and fill in the blanks; check your answers. Pronounce each number slowly and clearly with all its sounds.*

a. 50	e. 118	i. 913,048	m. _____	q. _____
b. 15	f. 430	j. 800,219	n. _____	r. _____
c. 16	g. 2,417	k. 12,717,870	o. _____	s. _____
d. 60	h. 15,380	l. 6,789,416	p. _____	t. _____

• *A second way to practice is to get a native English speaker to listen to you pronounce this list of numbers.*

6-04: READING

Directions: **Read the following paragraphs.**

PRONOUNCING FRACTIONS

3. Understanding and reading fractions

All the numbers you practiced on the previous page were **whole numbers**. They are used to count and describe whole things, like 3 pizzas, 45 minutes for lunch, and 5,000 bolts made today. Often, however, you will want to talk about **part** of something. For example the length of a part is two whole inches and part of another inch. You can use **fractions** to describe the length. There are two kinds of fractions: **bar fractions**, like 2 3/4, and **decimal fractions**, like 2.7500.

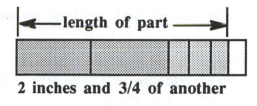

2 inches and 3/4 of another

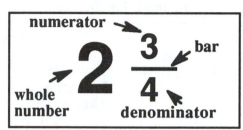

To give a bar fraction, you write two numbers: the number below the **bar**, called the **denominator**, tells how many parts were in the original whole; the number above the bar, called the **numerator**, tells you how many parts you actually have. When you say **3/4**, you are saying that you now have 3 out of an original 4 parts. (You will see decimal fractions on the next page.)

The expression **2 3/4** is called a **mixed number** because it is made up of a whole number, 2, and a fraction, 3/4. When you read and say a mixed number, put an "and" between the 2 and the 3/4; read it as "two <u>and</u> three-fourths."

The denominators must be pronounced carefully; you will be using fractions of an inch, so you should practice saying *half, quarter, quarters, eighth, eighths, sixteenth, sixteenths, thirty-second, thirty-seconds, sixty-fourth, sixty-fourths*. Notice the frequent use of the *-TH* ending on the end of the denominators. You need to make the following sounds differently: *fifty, fifteen, one fiftieth, three fiftieths, one fifteenth, and three fifteenths*: Try it! 50, 15, 1/50, 3/50, 1/15, 3/15. The person listening to you should be able to hear clear differences in your pronunciation.

6-05: PRONOUNCING BAR FRACTIONS

Directions: **Practice by listening to these bar fractions and then repeating the pronunciation of each. Then listen, fill in the blanks, and pronounce items p to y. Then have a native English speaker listen to you say them again.**

a. 50	f. 37/50	k. 14 3/8	p. _____	u. _____
b. 15	g. 3/4	l. 6 11/16	q. _____	v. _____
c. 1/50	h. 3/32	m. 18 1/64	r. _____	w. _____
d. 3/50	i. 27/64	n. 3 1/2	s. _____	x. _____
e. 1/15	j. 5/8	o. 7 3/100	t. _____	y. _____

6-06: READING
Directions: *Read the following paragraphs.*

PRONOUNCING DECIMAL FRACTIONS

4. Understanding and reading decimal fractions

One way to look at a decimal fraction is to think of it as an **abbreviation** for a bar fraction; it is a shorter way to write a fraction. The quantity "three and fifteen hundredths" could be written 3 15/100 in bar fractions or 3.15 in decimal fractions; with the bar fraction, the bar and the denominator must be written; with the decimal fraction, the two places after the decimal point show the decimal amount is hundredths. Study the values on this chart:

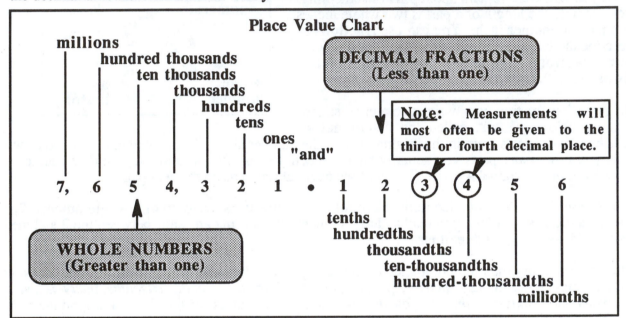

6-07: EXERCISE
Directions: *Write decimal fractions that are equal to these bar fractions:*

a. 3/10 = .3

b. 14/100 = .14

c. 65 8/1000 = 65.008

d. 250/10000 = .0250

e. 59 35/100 = _____

f. 28 5/100 = _____

g. 875/1000 = _____

h. 3 125/1000 = _____

i. 1/10000 = _____

6-08: PRONOUNCING DECIMAL FRACTIONS
Directions: *Practice by listening to these decimals and then repeating the pronunciation of each. Also fill in the blanks.*

a. .500

b. 1.55

c. 3.875

d. .0125

e. .9

f. 4.75

g. 614.5

h. 8.001

i. 29.1115

j. .065

k. _____

l. _____

m. _____

n. _____

o. _____

p. _____

q. _____

r. _____

s. _____

t. _____

6-09: READING
Directions: Read the following paragraphs.

LEARNING OTHER MATH-RELATED WORDS

5. Words used to talk about doing math problems

The goal of these pages is to help you with the language of math. Here are some useful words when you are doing problems:

a. It is possible to add, subtract, multiply and divide whole numbers, bar fractions, and decimal fractions. **Adding, subtracting, multiplying** and **dividing** are the four basic **operations** of mathematics.

b. Look at the numbers in these sample problems; learn the names for the parts of the problems:

```
          sum or total
25 + 15 = 40

          difference
17 - 3 = 14
```

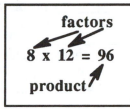

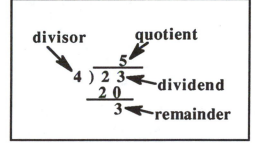

sum or **total** = the answer in an addition problem
difference = the answer in a subtraction problem
product = the answer in a multiplication problem
factors = the numbers which give you the product
        when multiplied together
dividend = the number to be divided
divisor = the number doing the dividing (How many
        times is the divisor contained in the dividend?)
quotient = the answer in a division problem
remainder = the amount left after the dividing has been done

Mathematical Operations

c. Some bar fractions are equivalent to other bar fractions. When you get an answer, you should check to see that it is in its **lowest terms,** by **reducing** it. For example, 4/8 is not in its lowest terms, because the top and the bottom can both be divided evenly by 4; when you divide, you get 1/2; you have reduced the bar fraction to its lowest terms; it can not be reduced any further. Some answers cannot be reduced, because there is no number which will divide evenly into the top and the bottom; for example, 7/10, 13/20, and 11/16 cannot be reduced.

6-10: EXERCISE
Directions: Read the statements. Circle T for "true," and F for "false."

T   F   1. The four operations are multiplying, dividing, adding, and subtracting.

T   F   2. In the problem 18 1/2 ÷ 3 3/4, the 3 3/4 is the dividend.

T   F   3. All of these bar fractions can be reduced: 5/15, 10/12, 8/16, and 25/100.

T   F   4. If you divide both the 12 and the 18 in 12/18 by 2, you have lowest terms.

6-11: CONVERSATION

Directions:
Lucy Garcia and Omar Harris are studying math together. Listen to their conversation and fill in the missing numbers or words.

Lucy: We've learned a lot of vocabulary for doing math _____. Shall we practice?

Omar: Sure, Lucy. I'd like to _____ the words and also do a few problems. Go ahead and ask me.

Lucy: Okay, Omar. What do you call the answer in a _____ problem?

Omar: The answer in a times problem is called the _____. Now can you tell me the answer in a _____ problem?

Lucy: Yes, the answer is called the _____. While we're on division, what do you call the two _____ that you start with?

Omar: The number to be divided is called the _____ and the number that's doing the dividing is called the _____. Lucy, ask me a few math problems that I can do in my _____.

Lucy: Give me the _____ of 7, 8, and ____. Then give me the _____ between 30 and _____.

Omar: The sum is the answer in an _____ problem, so I'll add 7, 8, and 9. Let's see, that would be _____. And to find the _____ I will subtract, so the difference between 30 and 17 is 30 minus 17 which is _____. I've got it!

6-12: EXERCISE
Directions:
1. Write a decimal fraction equivalent for each bar fraction:

a. 5/100 = _____   c. 75/10000 = _____   e. 7 9/10 = _____

b. 125/1000 = _____   d. 42 35/100 = _____   f. 185/10000 = _____

2. Reduce these bar fractions, if possible:

a. 50/100 = _____   c. 8 125/1000 = _____   e. 10/15 = _____

b. 24/32 = _____   d. 35 42/49 = _____   f. 5 17/20 = _____

3. Read these decimal fractions; write them as bar fractions; reduce, if possible.

a. .12 = _____   d. 12.500 = _____

b. 4.020 = _____   e. 25.04 = _____

c. .117 = _____   f. 52.625 = _____

6-13: VOCABULARY LIST B
Directions: *Study the vocabulary.* *Write the missing words in the blank spaces.*

1. degree

a unit of measure for angles and arcs. One degree = 1/360 of the circumference of a circle. *Example:* One _____ can be divided into 60 minutes.

1 degree = 60 minutes

2. plane

a flat surface in which plane figures, like circles, triangles, and squares, can be drawn. *Example:* A _____ can be represented by the surface of a flat piece of paper.

3. constant

a quantity that always has the same value. *Example:* The length of the diameter of a circle will always divide into the length of the circumference about 3.14 times; that value of 3.14 is a _____.

4. adjacent

lying next to something else. *Example:* In the triangle **ABC** side **a** is opposite **angle A** and **side b** is _____ to **angle A.**

5. Pythagorean theorem

a formula in mathematics with which the third side of a right triangle can be computed, if the other two sides are known. *Example:* The _____ was named after Pythagoras, a Greek mathematician from the 6th century BC.

6. to square

to multiply a number by itself. *Example:* When you _____ the number 12, you get the answer 144; you compute the square of 12 like this: 12 x 12 = 144.

7. square root

a number which when multiplied by itself, gives you the number you start with. *Example:* The _____ of 25 is 5. Checking: 5 x 5 = 25.

8. ratio

a comparison of one quantity to another; it can be expressed as a bar fraction. *Example:* The _____ of good parts to scrap on this shift is 350 to 1; scrap to good parts is 1 to 350 or 1/350.

9. function

a quantity whose value depends on another quantity. *Example:* The final price for a new house is a _____ of the cost of labor and materials.

10. key pad

the rows of keys on a calculator; the keys have numbers and functions printed on them. *Example:* He uses the _____ to enter the numbers with which he wants to add, subtract, multiply and do other operations.

key pad

calculator

(San Diego CCD)

57

6-14: READING
Directions: Read the following paragraphs.

6. Words used to talk about geometry problems

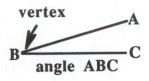

In the first drawing at the right, two lines, **AB** and **BC**, are the **arms** of an **angle**, **ABC**. The point where the two arms join, **B**, is called the **vertex** of the angle. (These ideas come from **geometry**, a branch of math that studies, points, lines, angles and planes.)

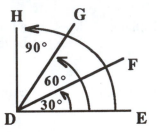

The arms of the angle can be opened up wider and wider; the amount of the angle's opening is measured in degrees. The second drawing shows various common angles and their measurements in degrees: **FDE = 30°**; **GDE = 60°**; **HDE = 90°**; these are some commonly-used angles. The name **right angle** is given to any angle of 90°.

As the arms of the angle continues to open, the size of the angle, as shown in the drawings below, goes past **90°** to **120°** (angle **IJK**), to **180°** (a straight line, angle **LMN**) to **270°** (angle **OPQ**), and finally completes a full circle at **360°** (angle **RST**); there are **360°** in any circle.

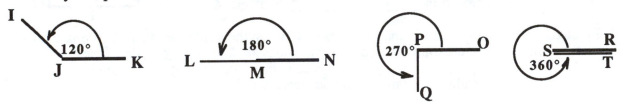

For more accurate angle measurements, a degree can be divided into 60 minutes (abbreviated with ') and each minute can be divided into 60 seconds (abbreviated "); a measurement of **45° 17' 32"** would be read *forty-five degrees, seventeen minutes, thirty-two seconds.*

7. More words from geometry

A **triangle** is a closed shape made from three straight lines and three angles. A triangle is an example of a **plane figure**, and is called that because the three lines are drawn on a flat surface called a **plane**; actually, a plane exists only in the mind as an idea, the idea of a perfectly flat space with only two dimensions. We represent a plane in our three-dimensional world by things like a flat piece of paper with a triangle, or circle, or other drawing on it.

When you add up the degrees in the three angles of any triangle, you will always get **180°**; the triangle at the right shows a triangle of 90°, 60° and 30°; 90 + 60 + 30 = 180.

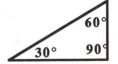

Any triangle which includes a right angle (an angle of 90°) is called a **right triangle**.

6-15: EXERCISE
Directions: Read the statements. Circle T for "true," and F for "false."

T   F   1. There are no angles larger than 90°.

T   F   2. An angle of 90° is called a right angle.

T   F   3. An angle can be measured in degrees, minutes, and seconds.

T   F   4. The three angles of any triangle always total 90°.

6-16: READING
Directions: *Read the following definitions.*

8. The parts of a circle
Another plane figure used frequently in the machining field is the **circle**. Study the parts of the circle, and learn these important words which will help you to talk about circles.

a. **center**   the point in the middle of the circle.

b. **circumference**   the outside edge of the circle; all points on the circumference are of equal distance from the center.

c. **radius**   A straight line drawn from the center to the circumference. (The plural of radius is **radii**).

d. **diameter**   A straight line drawn through the center which joins two points on the circumference. The diameter is related to the radius in this way: **d = 2r**, the diameter is 2 x the radius.

e. **chord**   A straight line which joins two points on the circumference but does not pass through the center.

f. **arc**   Any part of a curved line, especially a part of the circumference. An arc is measured by the angle that encloses it. If angle **ABC = 45°**, then **arc AC = 45°**.

g. **pi (π)**   When the distance around the circumference of any circle is divided by the diameter of that circle, the answer is always about 3 1/7; if the circle is larger, the diameter is longer, but the relationship is constant: the circumference is always about 3 1/7 times the length of the diameter. This number, a **constant**, is called pi, a Greek letter written π. More precise values for π can be reached: **3.1416** is a good approximation for π.

h. **C = dπ**   A **formula** can be written to express the relationship of the diameter to the circumference through the constant, π: **C = dπ**. Here C is the length of the circumference and d is the length of the diameter.

6-17: EXERCISE
Directions: *For each statement, circle T for "true" or F for "false."*

T   F   1. The circumference of a circle whose diameter is 7 inches will be about 22 inches.

T   F   2. An arc of 37° 3' 56" will be measured by an enclosing angle of 37° 3' 56".

T   F   3. The length of the radius is twice the length of the diameter.

T   F   4. All the radii of a circle are *equidistant* (same distance) from the center of that circle.

6-18: EXERCISE

Directions: *Match the letters with the correct numbers .*

_____ 1. C = dπ

_____ 2. plane

_____ 3. diameter

_____ 4. vertex

_____ 5. 3.1416

_____ 6. circumference

_____ 7. 60

_____ 8. radius

_____ 9. right angle

_____ 10. center

a. any angle equal to 90°

b. the distance around the outside edge of a circle

c. the point in the middle of the circle

d. the number of minutes in one degree of angle measurement

e. a flat two-dimensional surface which exists as an idea

f. any straight line connecting the center with the circumference

g. the value of pi given to the ten-thousandth place

h. the formula for finding the circumference of a circle

i the point where the two arms of an angle meet

j. a straight line passing through the center and connecting two points on the circumference

6-19: "SHOP TALK"   *USING NUMBERS AND MATH TERMINOLOGY*

Directions: *Some important instructions and conversations in the machine shop use numbers and math terminology. Listen to Le's questions & statements; then listen to three replies from Omar. Circle the best answer. Listen again. Practice your pronunciation.*

Le Tran's questions	*Omar Harris's replies*		
1. Which of these three values for π is more accurate?	A	B	C
2. How many degrees are in a right angle?	A	B	C
3. How many seconds are in a degree?	A	B	C
4. How many degrees are there in a circle?	A	B	C
5. If the radius is 3.750 in., how long is the diameter?	A	B	C
6. I want to bore a 2.250-inch hole in this workpiece with a tolerance of ±.003. What's the largest the hole could be?	A	B	C
7. I need to mark off an arc of 55° 12' 13" on the circumference of this circular workpiece. How do I do that?	A	B	C
8. In a right triangle with one angle equal to 65°, what is the size of the other angle which is not the right angle?	A	B	C
9. In the formula C = πd, which letter is a constant?	A	B	C

6-20: READING
Directions: *Read the following definitions.*

9. Some other formulas and abbreviations:

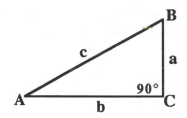

a. The Pythagorean theorem: Look at the right triangle ABC with angles A, B, and C, and with sides a, b, and c. The side opposite the right angle, side **c**, is always the longest side. It has a special name, the **hypotenuse**. For angle A, side a is called the **opposite side**, and side b is called the **adjacent side**. These names will be useful for the study of trigonometry.

The Pythagorean theorem says that the square of the hypotenuse is equal to the sum of the squares of the other two sides. Written in a formula, this is: $c^2 = a^2 + b^2$. When you square a number, you multiply it times itself; for example, $5^2 = 5 \times 5 = 25$ and $1.2^2 = 1.2 \times 1.2 = 1.44$. The opposite of squaring a number is taking a **square root**, shown by the sign $\sqrt{}$; to find a square root of a number, think of a number which, when multiplied by itself will give you the number; for example, $\sqrt{16} = 4$ and $\sqrt{49} = 7$.

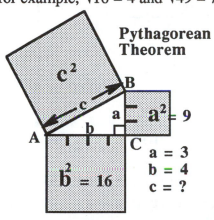

Pythagorean Theorem

$a^2 = 9$

a = 3
b = 4
c = ?

$b^2 = 16$

Example: In the right triangle at the left, side a = 3 and side b = 4; use Pythagorean theorem to find the length of c.

The formula says:	$c^2 = a^2 + b^2$
Substitute the values for **a** and **b** into the formula:	$c^2 = 3^2 + 4^2$
Squaring gives:	$c^2 = 9 + 16$
Adding gives:	$c^2 = 25$
Finding the square root of both sides:	$\sqrt{c^2} = \sqrt{25}$
	$c = 5$

b. Revolutions-per-minute (RPM) formulas: The machinist must often compute the speed at which to run a machine, in order to use a cutting or drilling tool with maximum efficiency and minimum wear. The following formula is a useful one: $RPM = \dfrac{4CS}{D}$

in which **D** = the diameter of the workpiece (round work on a lathe) or the diameter of the cutting tool (drill bit, milling cutter) and in which **CS** (the cutting speed given in feet per minute) = a value taken from a chart which recommends a certain number of surface feet per minute (**SFPM**); such a chart is given on page 277 of the textbook. This formula will be used later when you use the machines; for right now, memorize the formula.

c. Some common abbreviations

dia. = diameter	ID = inside diameter	in. = inch	m = meter
cham. = chamfer	OD = outside diameter	ft. = foot	dm = decimeter
dim. = dimension	PD = pitch diameter	max. = maximum	cm = centimeter
tol. = tolerance	TPI = threads per inch	min. = minimum	mm = millimeter

In the textbook, the inside of the front cover has a number of useful abbreviations and formulas. Take a look at it.

61

6-21: EXERCISE
Directions: Write short answers for these questions.

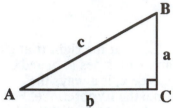

**Right triangle: the 90°
angle is indicated by
the square box at C.**

1. What is side **c** called? It's the _____

2. For angle A, what is side **a** called?
   It's called the _____ side.

3. For angle A, what is side **b** called?
   It's called the _____ side.

4. For angle B, what is side **a** called?
   It's called the _____ side.

5. For angle B, what is side **b** called? It's called the _____ side.

6. Write the formula for the Pythagorean theorem: _____

7. Square the following numbers: a. 3^2 = _____   c. 7^2 = _____
   b. 2.1^2 = _____   d. 5.3^2 = _____

8. Find the square root of the following numbers:
   a. $\sqrt{81}$ = _____   c. $\sqrt{4}$ = _____
   b. $\sqrt{100}$ = _____   d. $\sqrt{1/4}$ = _____

6-22: DICTATION

*Directions: Listen to this story and fill in the blanks with numbers
and with abbreviations for the words that are missing.*

Last week I made a part from bar stock. It was _____ long, _____
wide, and _____ thick. At one end of the work I drilled a hole with a
_____; the center of this hole was located _____ from one end of
the part. At the opposite end of the work, I drilled a blind hole which was _____
deep; the _____ of the hole was _____. I threaded the hole so it would take
a stud with a _____ of _____, a _____ of _____, and _____.
I then put a _____ on all edges of the part.

6-23: MATH NOTE

> The **Pythagorean theorem** is used to find the length of the third side of a triangle,
> when you already know the lengths of the other two sides. It doesn't matter which side
> of the triangle is missing; it can be found. The formula shown on the previous page is
> for finding **c** when a and b are known. Here are two formulas for the other two sides:
>
> **For finding a, when b and c are known: $a^2 = c^2 - b^2$**
> **For finding b, when a and c are known: $b^2 = c^2 - a^2$**

6-24: READING
Directions: *Read the following paragraphs.*

10. Some ideas about trigonometry

a. The definition: Trigonometry *is a branch of mathematics that studies the measurement of triangles.* It will be very helpful in computing important lengths and angles in machining work.

Ratio of Men to Women

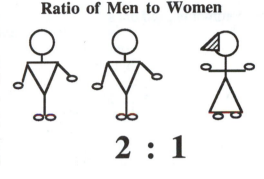

b. **What is a ratio?** Before going into trigonometry it is important to understand the idea of a ratio. A **ratio** *is the comparison of two quantities*; for example, in the shop there are 10 men and 5 women at work. In this example there is a ratio of 10 men to 5 women. This ratio can be written as a fraction 10/5 or 10 : 5 (which is read "10 to 5."); these expressions can be reduced: 10/5 = 2/1 or 10 : 5 = 2 : 1. There is always a second ratio if you reverse the comparison: 5 women to 10 men is 5/10 = 1/2, or 5 : 10 = 1 : 2, or .5 :1.

c. **What is a function?** A **function** *is one number value which depends on another number value.* For example, in the formula C = dπ, the value of C is a function of d: if d increases, C will get bigger; if d decreases, C will get smaller.

d. **The sine function:** Look again at the letters on the right triangle **ABC**: The angles of the triangle are given by the capital letters **A**, **B**, and **C**; the sides are given by the small letters **a**, **b**, and **c** (the hypotenuse). Below are three right triangles: they all have a hypotenuse of 4 in.; **angle A** changes from one picture to the next; **angle A** = 30°, then 20°, then 10°. As **angle A** gets smaller, side **a**, the side opposite, gets smaller: **a** = 2, 1 3/8, and 22/32. The decimal equivalents of these last three values are: 2 = 2.000, 1 3/8 = 1.375, and 22/32 = .71875.

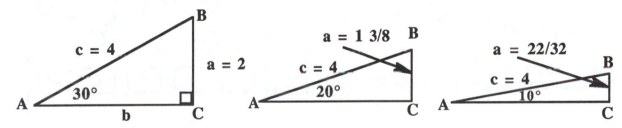

Now form a ratio using the opposite side over the hypotenuse, and call it **sine A**. (sin = sine)

So **sin A** = $\dfrac{\textbf{opposite side}}{\textbf{hypotenuse}} = \dfrac{a}{c}$ So using the measurements from above, substitute these values into the sine definition:

sin 30° = $\dfrac{2.000}{4.000}$ = .5000   sin 20° = $\dfrac{1.375}{4.000}$ = .3438   sin 10° = $\dfrac{.71875}{4.000}$ = .1797

These values can be arranged into a table, as shown on the next page. For each angle there will be a sine ratio, comparing the length of the side opposite the angle with the length of the hypotenuse. As the angle gets bigger, the sine ratio increases. Because one value depends on the changing of the other value, the sine ratio can be called the **sine function**.

10. Some ideas about trigonometry (continued)

e. **Other trigonometric functions:** Different combinations of the sides a, b, and c can be made to form other trigonometric functions. Here are the most commonly used ones:

		Abbreviation
sine A =	$\dfrac{a}{c}$ = $\dfrac{\text{opposite}}{\text{hypotenuse}}$	sin
cosine A =	$\dfrac{b}{c}$ = $\dfrac{\text{adjacent}}{\text{hypotenuse}}$	cos
tangent A =	$\dfrac{a}{b}$ = $\dfrac{\text{opposite}}{\text{adjacent}}$	tan

Angle	SIN
10°	.1797
20°	.3438
30°	.5000

A way to remember these is the sentence:
"Oscar Has A Hat On Always."
Oscar Has = Opposite / Hypotenuse = sin
A Hat = Adjacent / Hypotenuse = cos
On Always = Opposite / Adjacent = tan

f. **Functions in tables and calculators:** There are tables which give lists of all the functions for all angles found in right triangles; some tables even give values for the divisions of a degree: the minutes and the seconds. The number of decimal places given in the table can vary; there are usually 4 or 5 places. The **sample table** on the next page gives the angles in whole degrees and the function values to four places.

Here is a picture of a scientific calculator that has the trig functions available when the machinist presses its keys.

Look at the **trig function keys.** The general rule is: to get the trig function for an angle, enter the degrees of the angle, then press the function key (the three trig functions are marked with arrows and a dot.) For example, what is the sine of 45°? First press 4 and 5 on the number pad; then press the **sin key.** The result is .707106781 which can be shortened to **.7071.**

Now go to the trig table on the next page, and see if that's the number for sin 45°.

Look at a second example: Find tan 58°. Do this: Press 5 and 8; then press the **tan key.** You get 1.600334529 which you can be shortened to **1.6003.** For angles over 45° use the headings at the bottom of the page.

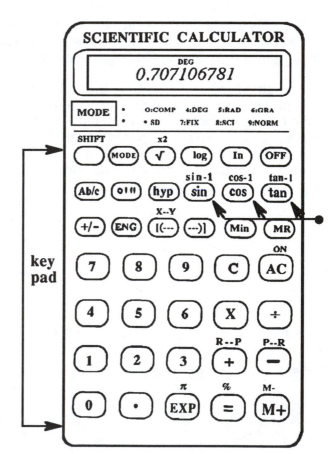

scientific calculator with key pad
(SDCCD)

Table of Trigonometric Functions

ANGLE	SIN	COS	TAN	COT	
0°	.0000	1.0000	.0000	∞	90°
1	.0175	.9998	.0175	57.2900	89
2	.0349	.9994	.0349	28.6363	88
3	.0523	.9986	.0524	19.0811	87
4	.0698	.9976	.0699	14.3007	86
5°	.0872	.9962	.0875	11.4301	85°
6	.1045	.9945	.1051	9.5144	84
7	.1219	.9925	.1228	8.1433	83
8	.1392	.9903	.1405	7.1154	82
9	.1564	.9877	.1584	6.3138	81
10°	.1736	.9848	.1763	5.6713	80°
11	.1908	.9816	.1944	5.1446	79
12	.2079	.9781	.2126	4.7046	78
13	.2250	.9744	.2309	4.3315	77
14	.2419	.9703	.2493	4.0108	76
15°	.2588	.9659	.2679	3.7321	75°
16	.2756	.9613	.2867	3.4874	74
17	.2924	.9563	.3057	3.2709	73
18	.3090	.9511	.3249	3.0777	72
19	.3256	.9455	.3443	2.9042	71
20°	.3420	.9397	.3640	2.7475	70°
21	.3584	.9336	.3839	2.6051	69
22	.3746	.9272	.4040	2.4751	68
23	.3907	.9205	.4245	2.3559	67
24	.4067	.9135	.4452	2.2460	66
25°	.4226	.9063	.4663	2.1445	65°
26	.4384	.8988	.4877	2.0503	64
27	.4540	.8910	.5095	1.9626	63
28	.4695	.8829	.5317	1.8807	62
29	.4848	.8746	.5543	1.8040	61
30°	.5000	.8660	.5774	1.7321	60°
31	.5150	.8572	.6009	1.6643	59
32	.5299	.8480	.6249	1.6003	58
33	.5446	.8387	.6494	1.5399	57
34	.5592	.8290	.6745	1.4826	56
35°	.5736	.8192	.7002	1.4281	55°
36	.5878	.8090	.7265	1.3764	54
37	.6018	.7986	.7536	1.3270	53
38	.6157	.7880	.7813	1.2799	52
39	.6293	.7771	.8098	1.2349	51
40°	.6428	.7660	.8391	1.1918	50°
41	.6561	.7547	.8693	1.1504	49
42	.6691	.7431	.9004	1.1106	48
43	.6820	.7314	.9325	1.0724	47
44	.6947	.7193	.9657	1.0355	46
45°	.7071	.7071	1.0000	1.0000	45°
	COS	SIN	COT	TAN	ANGLE

Directions: *Listen to the lecture that Omar recorded in class while Al Lopez, the teacher, was speaking. Fill in the blanks in Omar's notes with what you hear. Listen again and pronounce the words.*

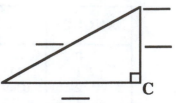

Here is a chart with some _____ - _____ solution _____ like the one on page _____ of the textbook. These formulas can be used to solve _____ problems.

angle A	side a	side b	side c
_____ = $\dfrac{a}{c}$	a = c x _____		_____ = $\dfrac{a}{\sin A}$
___ A = $\dfrac{b}{c}$		b = c x _____	c = $\dfrac{}{\cos A}$
_____ = $\dfrac{a}{b}$	a = _____ x tan A	b = $\dfrac{a}{\rule{1cm}{0.4pt}}$	

What does the chart mean? It tells you what formula to use. Choosing a _____ will depend on what parts of the right triangle you are given.

The first formula under "angle A" is the definition of sin A; under "side a," is a formula for finding side a when you are given the length of the hypotenuse (c) and angle A. Here's a sample problem:

Sample Problems (to nearest .001)

1. **Given:**
    length of c = _____
    and A = 35°
    Find:
    length of **a**

    Formula from chart above:
    a = c x sin A
    a = 12 x sin 35°
    a = 12 x .5736 (from trig table)
    a = _____

2. **Given:**
    length of a = _____
    and A = 30°
    Find:
    length of **c**

    Formula from chart above:
    c = $\dfrac{a}{\sin A}$ = $\dfrac{7.125}{\rule{1cm}{0.4pt}}$
    c = 7.125 ÷ .5 = _____
    c = _____

3. **Given:**
    length of a = _____
    and length of c = _____
    Find:
    length of **b**

    Pythagorean _____ :
    $b^2 = c^2 - a^2$
    $b^2 = 10^2 - 6^2$
    $b^2 = 100 - 36$ = _____
    $\sqrt{b^2} = \sqrt{64}$
    b = _____

6-28: MATHEMATICS TERMINOLOGY TEST

A. 🔲 *Listen and write what you hear.*

1. Write the whole numbers that you hear:

    a. _____   c. _____   e. _____

    b. _____   d. _____   f. _____

2. Write the bar fractions and mixed fractions that you hear:

    a. _____   c. _____   e. _____

    b. _____   d. _____   f. _____

3. Write the decimal fractions and mixed decimals that you hear:

    a. _____   d. _____   g. _____

    b. _____   e. _____   h. _____

    c. _____   f. _____   i. _____

4. Write the serial numbers:

    a. _____   b. _____   c. _____

5. Write the angle sizes you hear:

    a. _____   b. _____   c. _____

B. *Read the sentences below. Write the correct words or values in the spaces.*

1. Adding, subtracting, multiplying and dividing are the four basic mathematical _____.

2. Changing 8/16 to 1/2 is called _____.

3. The answer in a multiplication problem is called the _____.

4. The answer in a subtraction problem is called the _____.

5. In the fraction 17/32, the 17 is called the _____.

6. In the fraction 34 5/8, the 8 is called the _____.

7. In the problem 34 11/16 ÷ 2 1/2, the 2 1/2 is called the _____.

8. In the problem $14\,\overline{)\,343}$, the 343 is called the _____.

9. Write down the total number of degrees in each of these three triangles:

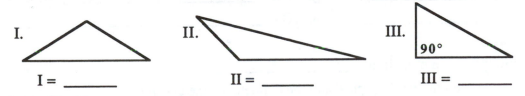

    I.   II.   III.   90°

    I = _____   II = _____   III = _____

10. What kind of triangle is triangle III? It's a _____.

67

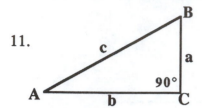

11. In triangle **ABC**, what is side **c** called?

It's called the _____.

12. In triangle **ABC**, for angle **A,** what is side **b** called? It's called the _____.

13. In triangle **ABC**, if you divide the length of side **a** by the length of side **c**, you will get a value for a trigonometric function known as the _____ of angle **A**.

14. Here are some names used to talk about problems dealing with circles. Write the names in the blanks.

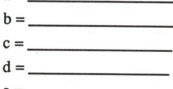

a = _____

b = _____

c = _____

d = _____

e = _____

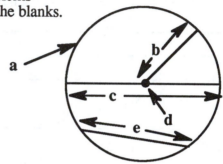

15. The value, **3.1416**, can be given by the Greek letter _____.

C. Abbreviations and formulas

1. Write an abbreviation for each of these words or expressions:

   a. inch = _____

   b. foot = _____

   c. 25 degrees = _____

   d. 4 centimeters = _____

   e. 25.3 millimeters = _____

   f. the sine of angle B = _____

   g. the cosine of angle A = _____

   h. the tangent of angle B = _____

2. Write out the complete word(s) for each of these abbreviations:

   a. dia. = _____

   b. ID = _____

   c. min. = _____

   d. CS = _____

   e. dim. = _____

   f. max. = _____

   g. RPM = _____

   h. OD = _____

3. Write the formulas for:

   a. the circumference of a circle: _____

   b. the RPM, given the cutting speed and the diameter of the cutting tool: _____

   c. the Pythagorean theorem, for a right triangle with sides a, b, and c: _____

Unit 7: *MEASURING INSTRUMENTS*

ASSIGNMENT: This unit covers material from the textbook, *Machine Tool Practices,* in Section C pages 83 to 193. Study that section, especially the pages referred to in this unit.

OBJECTIVES for this unit:
You should be able to:
1. Read and pronounce correctly the graduations shown on some common steel rules.
2. Read and pronounce correctly the readings from an outside micrometer and a vernier caliper.
3. Identify and say correctly the names of some other common measuring instruments.

7-01: VOCABULARY LIST A
Directions: Study the vocabulary. Write the missing words in the blank spaces.

1. standard — something established as a basis of comparison in measuring anything. *Example:* The inch is a _____ of measurement to which most American-made parts can be compared.

2. interchangeable — made the same so that one part can be used anywhere in place of another. *Example:* If parts made in the USA are _____ , they can be used in other parts of the world.

3. international — concerned with the relations among nations; something for the use of all nations. *Example:* _____ standard sizes are agreed upon by many nations and are used so that parts will be interchangeable.

4. instrument — a carefully made tool that can be used for fine measurements. *Example:* This _____ is a micrometer; it can be used to check dimensions given in thousandths.

an outside micrometer

5. to stamp — to imprint a mark, design, letter or number into metal by hitting the metal forcibly with a form for that mark, design, letter or number. *Example:* Fred wants _____ his name on all his new tools.

6. to discriminate — to divide a basic unit, like the inch, into smaller units which can be used to measure lengths with the accuracy required by the job. *Example:* A steel rule with 64ths can be used _____ a length given in 32nds.

7. to calibrate — to compare a measuring instrument to a known standard of measurement and adjust it to conform to that standard. *Example:* Lucy tries _____ her measuring instruments often.

8. to subdivide — to divide a unit into smaller units and then divide those units again. *Example:* The makers of steel rules are able _____ the inch into smaller and smaller units.

9. graduation — a mark stamped on a measuring device to show smaller units. *Example:* The smallest _____s I can see on this rule are the 64ths.

69

Directions: Read the following paragraphs.

THE IMPORTANCE OF MEASUREMENT

1. The world is using standard measurements, interchangeable parts, and mass production.

Today the whole world is tied together by new means of communication and an increase of trade among all nations. Because parts made in the United States are sold in Mexico, Canada, Europe, Asia, Latin America, Africa, or anywhere in the world, these parts must be made to meet international standards.

Standards of quality have been set, and among these is the need to make parts that are of **standard size**. When parts are made within the limits of given tolerances, the parts will be **interchangeable**, that is, they can be used anywhere; a bolt made in the USA can be used to attach an engine made in Japan to a frame made in Eastern Europe. Because the sizes are standard, the markets for the parts will be large, and there is often a demand for **mass production**, making certain parts in very large numbers.

The world uses interchangeable parts.

2. Some measurements require the use of measuring instruments.

A machinist is mostly concerned with measuring the length of parts. **Length is the distance from one point to another along a line.** The size of a part is given on a print by the length of the dimensions of the part; other words like *width* and *depth* can be used to describe length.

The machinist must be able to measure parts correctly and make them within a given tolerance. Many measuring devices have been developed to help the machinist do that. The word **instrument** is used to speak of these special tools; the word tells us we are using a carefully made tool, not like a hammer or some other tools which are not so exactly made.

7-03: EXERCISE
Directions: Rewrite each statement below, changing it to a yes/no question using "does" or "do" for present, and "did" for past. This kind of question expects a "YES" or a "NO" for an answer. Look at these examples:

S: *Joe calibrates his instruments often.*   Q: *Does Joe calibrate his instruments often?*
S: *Marina compared the two lengths.*   Q: *Did Marina compare the two lengths?*

1. The USA mass produces many parts.   Q: _____

2. Beatriz measured the part carefully.   Q: _____

3. Al taught the class about accuracy.   Q: _____

4. The students understand discrimination.   Q: _____

5. Germany and Japan have quality standards.   Q: _____

6. The machinist chose his tools for the job.   Q: _____

7-04: READING
Directions: *Read the following paragraphs.*

SOME MEASUREMENT TERMS

Because measuring is so important, a whole science of measurement has been developed. It is called **metrology**. There are certain terms from metrology that are important for a student of machine tool technology to remember:

1. Accuracy: a machinist may need to verify that the size of a tool or a part is really what it says it is. For example, a drillbit may have a size stamped on it. The machinist needs to measure what its actual size is; the size may be different, if, for example, the drillbit has been sharpened several times. If the measured size of the tool matches the size stamped on it, the tool size is said to be **accurate**. The measured length of some part can also be said to be accurate if that length matches the stated length on a drawing, within the limits of a given tolerance. Inaccuracies can also occur if the machinist uses incorrect procedures for measuring or if the instruments are not accurate.

2. Precision: This term refers to how closely a particular measurement must match a desired size. There are many degrees of precision possible in measuring things. Any measurement made finer than one sixty-fourth of an inch or one-half millimeter is called **precision measurement**.

3. Reliability: The term "reliability" refers to the ability of a tool to take a measurement that is as precise as is called for by the situation. For example, a yard stick is good for measuring how much cloth is needed to make a pair of pants, but it would not be reliable for measuring the diameter of a tiny gear which is to go inside a watch.

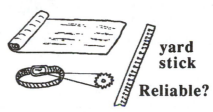

yard stick

Reliable?

4. Discrimination: This term refers to the degree to which a basic unit of length on a measuring instrument is divided into smaller units. You can buy a drugstore ruler that divides the inch into eighths, or you can use a machinist's steel rule that divides the inch into hundredths. If you want to measure a length that is 3.6 inches, you will use the steel rule instead of the drugstore ruler. The hundredths on the steel rule are more suited to your purpose than the eighths. Generally, measuring instruments should discriminate ten times finer than the smallest unit to be measured.

5. Calibration: Calibration is a process by which a measuring instrument is compared to a known standard of measurement and adjusted to conform to that standard. To achieve reliable results, the machinist must know that he/she has calibrated the measuring instrument to an appropriate standard or that the instrument has been calibrated in the metrology lab.

7-05: EXERCISE

[cassette icon] *Directions*: *Listen to the tape. Fill in the blanks with what you hear.*

Bill, a machinist, makes sure to send his measuring instruments regularly to the _____ lab to be_____. He also measures tools like drillbits before he uses them; he wants to make sure that they are_____. Often he deals with measurements that are less than one sixty-fourth of an inch, so he chooses an instrument for _____ measurement; in that way he can be sure of _____ results. He tries to choose instruments which have their smallest unit _____ times finer than the smallest unit he wants to measure. If he wants to measure a dimension given in hundredths of an inch, he chooses an instrument which will _____ to the nearest one thousandth.

7-06: READING
Directions: Read the following paragraphs.

UNITS OF MEASUREMENT: FRACTIONS OF AN INCH

Because measuring has been a requirement of daily life for most of human history, several systems of measurement have been developed. The two most important systems in the world today are the English system, which can be called **the inch system**, and the International Metric System which we will call **the metric system** and abbreviate as SI (for *Systeme International d'Unités*).

1. The Inch System:

The inch is the basic unit of the inch system. There are twelve inches in one foot and three feet in one yard. For machining purposes, the inch is broken down into smaller units named by bar fractions, such as 1/2, 3/4, 5/8, 7/16, 15/32, and 57/64; the inch can also be subdivided into smaller units named by decimal fractions, such as .5, .25, .625 and .0015.

The subdivisions of the unit inch shown by **bar fractions** are made by cutting the larger units into two pieces. For example, the first subdivision of the unit inch is one half. This is produced by cutting the unit inch into two pieces. Each piece is equal to 1/2 in. A long mark in the middle shows the 1/2 inch. A big "1" is used to show the whole inch.

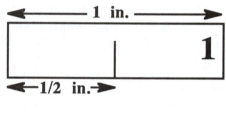

To produce the next smaller units the larger pieces are cut into two pieces again and again. Halves are cut into four fourths; fourths are cut into eight eighths; the eighths are cut into sixteenths. On a ruler, each time the unit is cut into smaller subdivisions, the line used to mark the new unit is shorter than the unit before it.

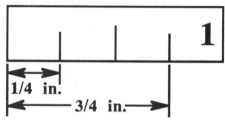

This subdividing could continue, but the eye is limited in being able to see down to about 1/64 of an inch. That is the finest division you will usually find. Some workers need a magnifying glass to read these small marks accurately.

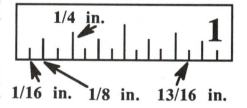

The inch can also be subdivided using **decimal fractions**; the unit inch can be divided into tenths, hundredths, and thousandths and placed on one or more edges of a measuring instrument like the steel rule. Decimal measurements as fine as ten-thousandths are sometimes required but they will not be visible to ordinary viewing. Some of the other instruments will be better for measuring those fine lengths than is the steel rule.

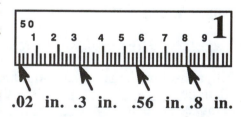

FRACTIONAL INCHES

Decimal fractions of an inch like these are often all read as thousandths: .020, .300, .560, and .800. The little 50 in the corner of the rule tells you there are fifty subdivisions of the inch shown; so, each short mark is worth .02 of an inch (50 x .020 = 1.000 inch.). Some rules will show the inch divided into 100 pieces, but the hundredths marks are difficult to read.

7-07: EXERCISE

 ___Directions:___ *Listen to the tape. Follow the verbal instructions.*

Part A: DIMENSIONS

Figure 7-01

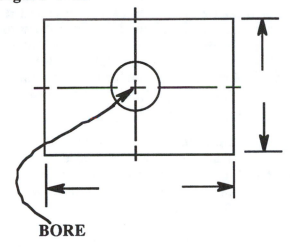

BORE

Figure 7-02

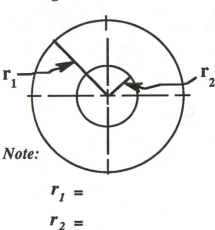

Note:

$r_1 =$

$r_2 =$

Part B: DECIMAL EQUIVALENTS

These are some bar fractions that are used _____ in the machine shop. It is useful to be able to change these fractions into their decimal _____ when needed. It will be good to _____ these bar fractions and their equivalents expressed as _____ fractions.

If you know these _____ by heart you will not have to look them up or use your _____ to figure them out.

1.	1/2	=	_____
2.	1/4	=	_____
3.	3/4	=	_____
4.	1/8	=	_____
5.	3/8	=	_____
6.	5/8	=	_____
7.	7/8	=	_____
8.	1/16	=	_____
9.	1/32	=	_____
10.	1/64	=	_____

Part C: CONVERSATION

Al: Let me say a bar _____, and you give me its decimal _____.

Lucy: Okay, I'm ready.

Al: Give me the decimal equivalents for 3/8, 1/16, and _____.

Lucy: Okay, 3/8 equals _____; 1/16 equals _____; and _____ equals _____.

7-08: READING
Directions: *Read the following paragraphs.*

UNITS OF MEASUREMENT: METRIC UNITS

2. The Metric System:

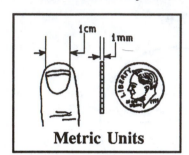
Metric Units

The world's other major system of measurement is the **metric system**. Its basic unit of measure is the **meter** (abbreviated **m**, with no period); a meter is equal to 39.37 inches. To form some useful subdivisions, the meter is divided into 100 **centimeters** (abbreviated **cm**) and into 1000 **millimeters** (abbreviated **mm**). A centimeter is about the width of a small fingernail, and the millimeter is about the thickness of a dime. The complete set of SI (metric system) units are given on page 101 of the textbook.

Most production of parts in the USA uses the inch system. Most of the other industrialized nations, like Germany and Japan, use the metric system. Sometimes there will be a need to convert measurements from one system to the other. Because the two systems have different histories, there are no easy equivalents between inch measurements and metric measurements. However, here are some conversion factors that you can memorize.

1 in. = 25.4 mm   **1 in. = 2.54 cm**   **1 mm = .03937 in.**

Sample conversion problems:

    a. Changing inches to metric units:
        (1) What is the length in millimeters of a dimension which is 4.750 in.?
        Answer: Given inches, to find millimeters, multiply the inches by 25.4.
                      4.750 x 25.4 = 120.65 mm

        (2) What is the length in centimeters of a radius which is .580 in.?
        Answer: Given inches, to find centimeters, multiply the inches by 2.54.
                      .588 x 2.54 = 1.473 cm

    b. Changing metric units to inches:
        (3) What is the length in inches of a tool edge which is 15 mm long?
        Answer: Given mm, to find inches, multiply the mm by .03937.
                      15 x .03937 = .591 in.

7-09: EXERCISE

Directions: *Listen to the tape. Fill in the blanks with what you hear.*

There are two systems of _____. The USA uses the _____ system, while most of the other industrialized _____ use _____. Two useful _____ units are the _____ , which is about the width of a little _____ , and the _____ , which is about the thickness of a _____. A few useful conversion formulas are: 1 in. = _____, 1 in. = _____ , and 1 mm = _____.

7-10: READING
Directions: *Read the following paragraphs.*

THE STEEL RULE

1. The steel rule and discrimination: Every machinist should own a steel rule. A common length for these rules is six inches; that size is easy to carry in a shirt pocket for easy use. Such a rule will be useful for measuring important dimensions. Different steel rules have different sets of graduation marks stamped on them.

The extent to which a unit of length has been divided is called discrimination. A rule which has been divided into 32 graduations is said to discriminate to 1/32 in. For common steel rules, the maximum discrimination possible is 1/64 in.; on a decimal inch steel rule, the maximum discrimination is .01 inch, and on a metric steel rule, the maximum discrimination is .5 mm. If you need to have discrimination at thousandths of an inch, other more sensitive measuring instruments would be better.

Here is an enlarged picture of the first inch of a steel rule. The graduations on one edge are in thirty-seconds, shown by the word "32NDS" stamped into the metal; the other edge shows sixty-fourths and has the word "64THS."

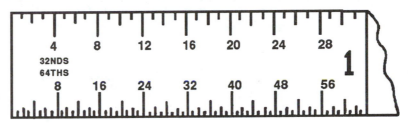

STEEL RULE with graduation marks for 32nds & 64ths

2. Reading the steel rule: Every new machinist needs practice in how to read a steel rule, so that he or she can make quick, accurate measurements. To begin practicing how to read the rule, look at the examples below.

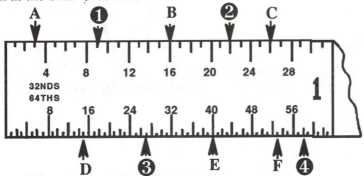

On the 32nds edge every fourth mark is labeled: 4, 8, 12, etc. A is located one mark before the 4, so it is read 3/32 inch. **B** is located at 16/32, but it will be read 1/2 when 16/32 is reduced. **C** is located two marks after the 24, so it marks the length 26/32 which is reduced to 13/16 inch.

On the 64ths edge every eighth graduation is labeled: 8, 16, 24...56. **D** is located one mark short of 16, so it is 15/64 inch. **E** is located at the 40 mark, so it is 40/64, which is reduced to lowest terms as 5/8 inch. **F** is located 5 marks after the 48 mark, so it is 53/64 which cannot be reduced.

7-11: EXERCISE
Directions: *Look again at the picture of the rule. Write values for:*

❶_____   ❷_____   ❸_____   ❹_____

7-12: VOCABULARY LIST B
Directions: *Study the vocabulary. Write the missing words in the blank spaces.*

1. micrometer ("mike")

an instrument for measuring lengths, diameters, and distances with a high degree of discrimination. *Example:* This inside _____ will discriminate to .001 inch.

inside micrometer

2. ratchet

ratchet

a device which allows turning the threaded parts of a mike until a snug fit is reached and then will continue to turn without adding pressure. *Example:* The _____on a micrometer allows for a snug fit around the part without damaging the interior threads of the mike.

3. recess

a hollow place, or a space moved back from the surface. *Example:* Sometimes it is difficult to measure the depth of a _____.

4. index

a pointer or indicator, like a needle on a dial. *Example:* The _____ line on the micrometer points to 9 on the thimble.

index line   **thimble**
(San Diego CCD)

5. caliper (or calipers)

an instrument with a pair of jaws or legs used to measure lengths. *Example:* Joe used a pair of _____ to measure the outside diameter of the pipe.

6. fixed

not moving, remaining in one place. *Exampl*e: A pair of calipers often has a pair of jaws, one of which is _____.

beam or bar

7. movable

able to be moved. *Example:* Rhonda has a caliper with one fixed jaw and one _____ jaw that will slide open to receive the object.

movable jaw

fixed jaw   **object**

8. beam (or bar)

the straight rod upon which the movable jaws slide and which shows the graduation marks of the main scale. *Example:* The rod with the main scale can be called the _____ or the _____ .

9. to coincide

to have the same position in space. *Example:* When the line on the vernier scale can be made to line up or _____ with a line on the main scale, it is called the " line of coincidence."

main scale

vernier scale

Lines coincide at this location

10. vernier

a short graduated scale that slides along a longer graduated instrument and is used to indicate fractional parts of graduations. *Example:* Rogelio found the line of coincidence on the _____ scale.

76

7-13: READING
Directions: Read the following paragraphs.

MICROMETER PARTS AND USES

Before studying micrometers, you may want to improve your use of the steel rule. You need to get a steel rule and practice with it. For now there are additional examples and practices on pages 109 to 112 in the textbook. Also study the different kinds of steel rules that are pictured on pages 105 to 108 in the textbook.

1. The main parts and common uses of micrometers

The micrometer is able to discriminate fine measurements more closely than the steel rule. A standard micrometer is able to discriminate to **.001** of an inch. There is also a **vernier** form of this tool which discriminates to **.0001** of an inch. There are three common types of micrometer, the **outside micrometer**, the **inside micrometer**, and the **depth micrometer**. You will study the use of these micrometers first, then you may want to look at the variety of other "special-purpose" micrometers explained on pages 125 to 128 of the textbook.

a. The outside micrometer

This instrument is called an **outside micrometer**, because it fits around the outside of the object which is being measured. Turning the **ratchet** at the end of the micrometer moves the **spindle** toward the fixed **anvil** or away from it.

For measuring, the micrometer is opened; then the anvil and spindle are tightened snugly around the object. The ratchet prevents the **threads** inside the micrometer from being damaged; it continues to turn, but not to tighten. The reading of the micrometer can be locked in place by turning the **lock nut**.

Outside Micrometer Parts

b. The inside micrometer

inside micrometer

The inside micrometer is used to measure the diameter of holes or other cavities in a part. The inside mike can be read directly or it can be transferred to be read by an outside micrometer or a steel rule. Like many micrometers these mikes often come as a set with a variety of lengths.

c. The depth micrometer

The depth micrometer is used to measure the depth of holes, grooves, or recesses. The depth mike uses a set of extension rods of different lengths. The set of rods should be kept with the original mike and not mixed with other mikes. The graduations are numbered in the direction opposite the outside mike. See textbook, p. 137-139.

depth micrometer

7-14: EXERCISE
Directions: Write short answers for each of these questions:

1. Write down the names for three different kinds of micrometers.

   a. _____   b. _____   c. _____

2. What amount of discrimination (in inches) does the ordinary micrometer have? _____

3. To what degree (in inches) does the vernier micrometer discriminate? _____

4. What does the depth micrometer use to reach to the bottom of a hole? _____

7-15: READING
Directions: Read the following material.

2. How to read an outside micrometer

Here are the steps, given in an example, for reading an outside micrometer; you will want to do them slowly and carefully at first; with practice you will do them more quickly and easily:

Steps	**Results**	

1. Read the largest visible number on the index line of the sleeve. Each is worth **.001 in.**   .100

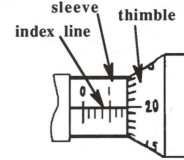

2. Count the number of sleeve marks past the visible number on the index line and multiply by .025: 2 x .025 = .050   .050

3. Count the number of marks past zero on the thimble and multiply by .001; the index line points to that reading: 20 x .001 = .020   + .020

Can you read this micrometer?
(San Diego CCD)

4. Add up the previous numbers; the total is the correct reading.   **Total = .170**

7-16: EXERCISE
Directions: Follow the steps and give the final reading for each of these micrometers.

A.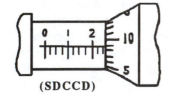
(SDCCD)

Step 1: _____
Step 2: _____
Step 3: _____
Total: _____

B.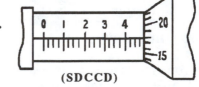
(SDCCD)

Step 1: _____
Step 2: _____
Step 3: _____
Total: _____

78

7-17: READING
Directions: Read the following paragraphs.

HOW TO READ A VERNIER MICROMETER

Look now at the sleeve and thimble of a vernier outside micrometer. Given here are the steps for reading this micrometer. The first three steps are the same as before. For the fourth step, look at the marks on the sleeve and find one that lines up with one of the marks on the thimble. The number from the vernier scale is read, not the thimble mark.

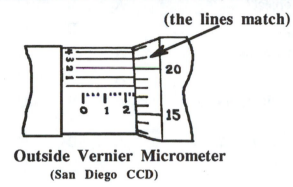

(the lines match)

Outside Vernier Micrometer
(San Diego CCD)

This micrometer with the vernier feature can be used for measurement of the **.0001** place. There are ten vernier markings on the micrometer which adds a tenth more after the thousandths place. This micrometer is sometimes called a "tenth mike."

Step 1: Visible sleeve number:	**.200"**
Step 2: Number of sleeve marks: 2 (.025) =	**.050**
Step 3: Number of thimble marks past 0:	**.018"**
Step 4: Find line on vernier that lines up with a thimble mark:	**.0003**
Step 5: Add up the results of Steps 1 to 4: Total =	**.2683"**

7-18: EXERCISE
Directions: Follow the steps and give the final reading for each of these micrometers.

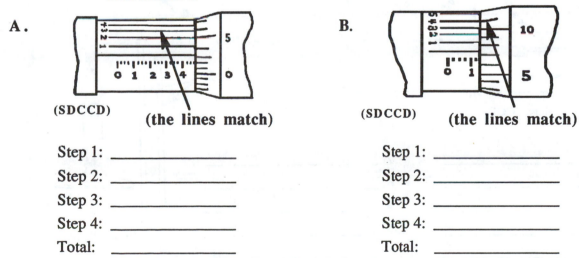

A.

(SDCCD)   (the lines match)

B.

(SDCCD)   (the lines match)

A.
Step 1: _____
Step 2: _____
Step 3: _____
Step 4: _____
Total: _____

B.
Step 1: _____
Step 2: _____
Step 3: _____
Step 4: _____
Total: _____

Note: For additional practice in reading micrometers, see page 135 (Fig. C-134a to C-134e) and pages 144-145 (Fig. C-154a to C-154e) in the textbook.

7-19: "SHOP TALK"   QUESTIONS and DIRECTIONS

*Directions: Listen to each instructor question. Then listen to three possible student answers. Circle A, B, or C for the **best** answer. Practice pronunciation as you repeat each question and answer.*

Instructor Question   **Student Answer**

1. Which parts of the outside mike go around the object?   A   B   C
2. Which part of the mike locks the reading in place?   A   B   C
3. What is the greatest degree of discrimination on a vernier mike? A   B   C
4. What is another name for a vernier micrometer?   A   B   C
5. Which instrument would you use to measure the diameter of a   A   B   C
   an automobile engine cylinder?
6. How do you read the final thousandth on an outside mike?   A   B   C
7. What's one way to measure the depth of a hole?   A   B   C

7-20: NOMENCLATURE EXERCISE:

Directions: Study the names of these instruments and their parts on the previous pages. Then listen to the names and write the them in the spaces below.

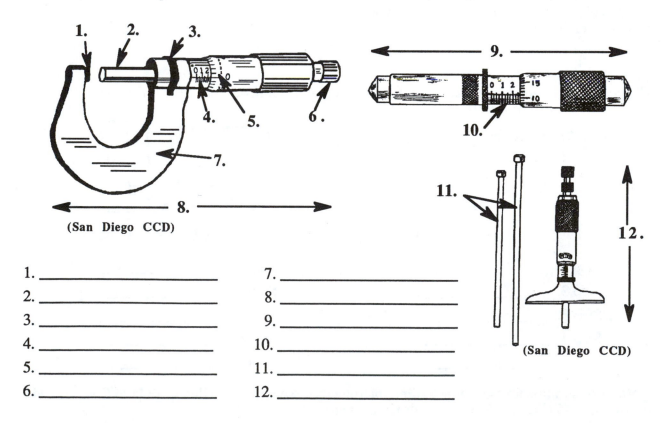

(San Diego CCD)

(San Diego CCD)

1. _____   7. _____
2. _____   8. _____
3. _____   9. _____
4. _____   10. _____
5. _____   11. _____
6. _____   12. _____

80

7-21: READING
Directions: Read the following paragraphs.

MEASURING WITH CALIPERS

(SDCCD)

OUTSIDE CALIPERS

Another group of measuring instruments are the family of **calipers**. A caliper always has **legs** or **jaws** which can be opened to hold the object to be measured.

1. Outside calipers: This simple caliper is shown at the right, with its **adjusting nut** for opening and closing the legs. After the length has been measured, the caliper is removed, and the length between the legs is measured directly with a steel rule or an inside micrometer. This is **indirect measurement** in which a direct measuring instrument is applied to a length taken earlier.

2. Vernier calipers: Another more-complicated type of caliper is the **vernier caliper**. It has a pair of **inside jaws**, for measuring inside spaces like holes, and a pair of **outside jaws** for measuring things like the outside diameter of a piece of pipe. Two of the jaws are **fixed jaws**, and two are **movable jaws** which slide along a **bar** or **beam**. The bar is marked with a **main scale** in whole inches, along with subdivisions of each inch into tenths.

The movable jaw is attached to a sliding **vernier scale**. A **clamping nut** allows the reading to be locked in place. An **adjusting screw** allows the caliper scales to be adjusted before use.

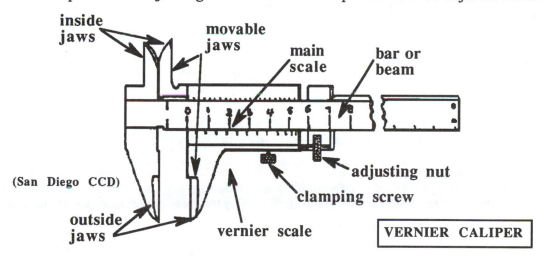

VERNIER CALIPER

NOTE about the vernier scale: The purpose of a vernier scale is **to increase the discrimination** of graduated-scale measuring tools, such as vernier micrometers and vernier calipers. A vernier system has a main scale and, next to it, a vernier scale.

The marks on the vernier scale are closer to each other than are the marks on the main scale, so that there is always one more graduation mark on the vernier scale than on the main scale. At some place along the two scales, a mark on the vernier scale will line up with a line on the main scale; that is called the **line of coincidence** and that is the extra fractional number to be added at the end of the answer.

7-22: EXERCISE

Directions: Study the scales on this vernier caliper. What is the reading shown? Then try giving the readings from the pictures in the textbook, pages 120-121, Figures C-88a to C-90b.

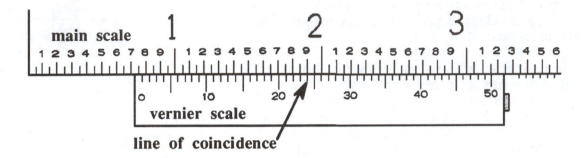

The main scale has three numbered inches showing in this drawing. **Each inch is divided into tenths** (.100, .200, etc.) which are also numbered, with smaller numbers. Each tenth of an inch is divided into two pieces by an unnumbered mark in the middle of each tenth; each mark on the main scale then is equal to half of a tenth (1/2 x .100 = .050). So between any two adjacent marks on the main scale there are 50 thousandths. There are 50 marks on the vernier scale that cover 49 marks on the main scale. In other examples, you may see 25 marks on the vernier scale covering 24 marks on a main scale each inch of which is divided into 4 pieces of .025 each.

The **zero mark** on the left-hand side of the vernier scale is the pointer to the reading on the main scale. The zero mark points, in this example, to a little beyond .700. The **line of coincidence** in this example is at 24 on the vernier scale. That 24 designates 24 thousandths to be added to the end of the reading. For the final two digits of the answer, remember to read the 24 from the vernier scale and not the 9 from the main scale. The final reading therefore is **.724 inch**.

7-23: READING

Directions: Continue reading the materials below on calipers.

3. Dial calipers: The **dial caliper** has most of the features of the vernier caliper, except it has a **dial** with 100 subdivision marks, each of which equals **.001**; thus its readings add the last two decimal places to the final reading given in thousandths. The tenths are read from the main scale which has actual inches with ten graduations.

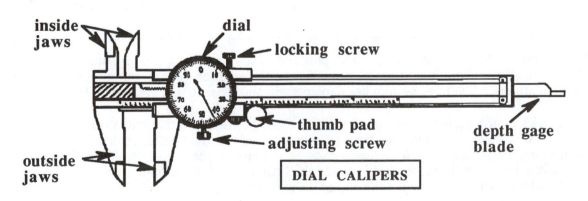

DIAL CALIPERS

7-24: NOMENCLATURE EXERCISE

Directions: *Study the names of these instruments and their parts on the previous pages. Then listen to the names and write the them in the spaces below.*

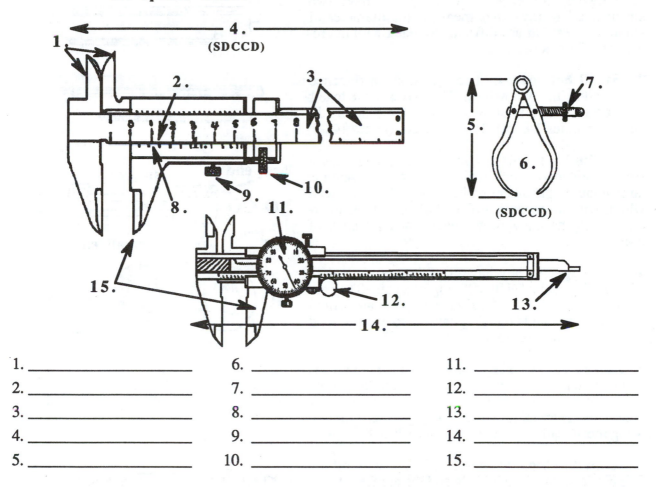

1. _____
2. _____
3. _____
4. _____
5. _____

6. _____
7. _____
8. _____
9. _____
10. _____

11. _____
12. _____
13. _____
14. _____
15. _____

7-25: EXERCISE

Directions: *Rewrite each statement below, changing it to a yes/no question using "does" or "do" for present, and "did" for past. This kind of question expects a "YES" or a "NO" for an answer. Look at these examples:*

S: *Those calipers have a depth gage.*   Q: ***Do those calipers have a depth gage?***
S: *Kim used a vernier caliper to measure the diameter.*   Q: ***Did Kim use a vernier caliper to measure the diameter?***

1. The vernier scale slides along the main scale.   Q: _____

2. I locked the reading in place with the clamping screw.   Q: _____

3. Outside calipers have two legs and an adjusting nut.   Q: _____

4. Danny inserted the depth gage blade into the hole.   Q: _____

5. The anvil and the spindle fit snugly around the part.   Q: _____

6. Carol uses extension rods on her depth micrometer.   Q: _____

7-26: OTHER MEASURING INSTRUMENT NOMENCLATURE

Directions: *Study these "other measuring instruments," and read about them in your textbook on the pages indicated.*

1. Telescoping gages: These gages have two spring-loaded arms that measure the diameter of smaller holes that are difficult to reach with micrometers. (Textbook, pages 147-8)

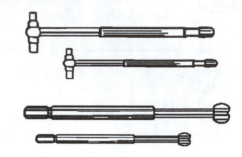

2. Small hole gages: Each gage has a flattened ball divided into two halves. The halves move together or apart when a knob is turned. For measuring diameter in holes 1/8 to 1/2 inch in size. (Pages 148-9)

3. Cylindrical plug gages: Each gage has a small "GO" end which will go into a hole which is large enough; the other end has a large "NO GO" end which will not receive it if the hole is outside tolerance. They have a variety of sizes for a quick test of hole diameters.

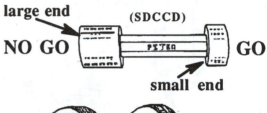

large end   (SDCCD)

NO GO   GO

small end

4. Ring gages: They have a variety of sizes with a "GO" ring and a "NO GO" ring as a pair. They are a quick check to see if cylindrical parts are within tolerance; the "NO GO" ring has a groove cut around it to identify it. (Textbook, pages 88-89)

GO   NO GO

(SDCCD)   groove

5. Thread gages: They come in a variety of sizes and each has a "GO" and a "NO GO" end for measuring internal threads and two threaded rings for measuring external threads. The gages are for measuring the size of any threaded parts. (Page 88)

GO   NO GO

6. Screw pitch gage: A set of blades with different size teeth on each blade and the number of "teeth per inch" stamped on each blade. The teeth of the correct pitch will fit into the threads of the threaded part being measured.

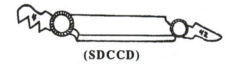

(SDCCD)

7-27: NOMENCLATURE EXERCISE

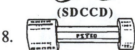 *Directions: Study the pictures and information from above. Then listen to the tape and follow the instructions.*

1. _____

2. _____

3. _____

4. _____

5. _____

6. _____

7. _____

(SDCCD)

8. _____

9. _____

10. _____

11. _____

84

Unit 8: *LAYOUT TOOLS & PROCEDURES*

ASSIGNMENT: **Read, study and complete pages 85-94 of this book. Then read in the textbook,** *Machine Tool Practices, SECTION E, Layout,* **pages 234-264; also read pages 152-185.**

OBJECTIVES for this unit:
You should be able to:
1. Tell the difference between semi-precision and precision layout.
2. Identify and pronounce correctly the nomenclature of some important tools for layout.
3. State the purposes of some layout tools.

8-01: VOCABULARY LIST A
Directions: **Study the vocabulary. Write the missing words in the blank spaces.**

1. to lay out   to measure and make marks on a workpiece to show the sizes of the part and its features. *Example:* Hans is planning _____ the features of the part on this piece of aluminum

2. operation   the use of a machine tool in any way in the making of a part. *Example:* Maria, the first _____ will be cutting off a workpiece from rough stock.

3. optical   related to equipment which uses light to make very fine measurements. *Example:* "The _____ comparator is used in the inspection of parts, cutting tools, and other measuring instruments." (Textbook, p. 97)

4. reference   a surface of known flatness or a point from which other lines and locations can be measured. *Example:* A surface plate provides a _____ surface from which to measure important features on a part.

5. granite   a very hard rock, gray to pink in color, from which some surface plates are made. *Example:* The _____ surface plate was made very smooth by a grinding process called *lapping*.

surface plate
(SDCCD)

6. layout dye   a quick drying liquid, usually blue, which is applied to a workpiece surface. Layout lines are then visible when a tool removes the dye. *Example:* This _____ is dry. Let's lay out the lines on the workpiece.

7. layer   a single thickness of some material. *Example:* A _____ of layout dye is first brushed over the surface of the workpiece

8. dent   a small hollow made in a surface by hitting it. *Example:* A hit with a hammer left a _____ in the workpiece.

9. bevel   an angle other than a 90° angle; a part or surface which runs at an angle to the rest of the part. *Example:* A chamfer is an example of a _____.

10. protractor   an instrument in the form of a half circle used for measuring and making angles. *Example:* I used a _____ to make an angle of 59°.

8-02: NOMENCLATURE

Directions: *Study the following pictures. Listen to the pronunciation of the names of the layout tools and accessories. Practice your pronunciation.*

NOMENCLATURE OF LAYOUT TOOLS--Part A

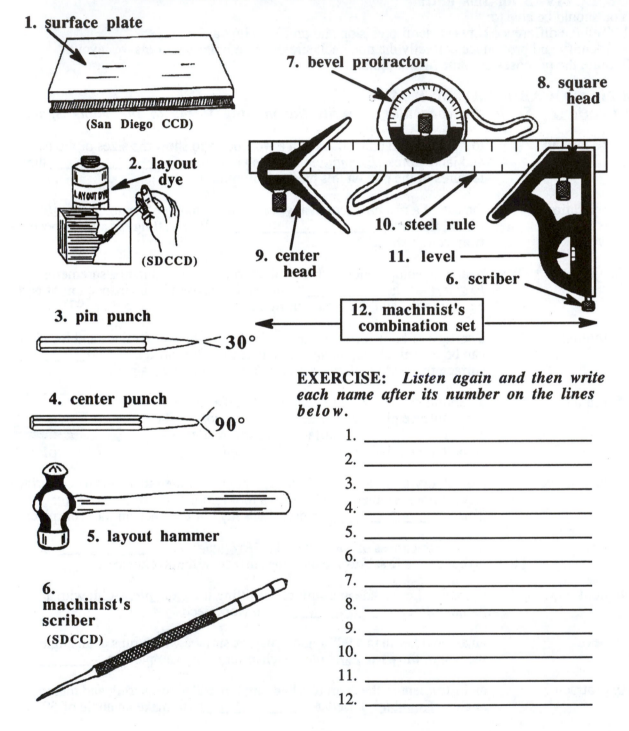

1. surface plate

(San Diego CCD)

2. layout dye

(SDCCD)

3. pin punch < 30°

4. center punch < 90°

5. layout hammer

6. machinist's scriber
(SDCCD)

7. bevel protractor

8. square head

9. center head

10. steel rule

11. level

6. scriber

12. machinist's combination set

EXERCISE: *Listen again and then write each name after its number on the lines below.*

1. _____
2. _____
3. _____
4. _____
5. _____
6. _____
7. _____
8. _____
9. _____
10. _____
11. _____
12. _____

Directions: Read the following paragraphs:

LAYOUT OPERATIONS

1. What is "layout"?

Layout is making marks on a workpiece to show the size of the part's features. For simple operations like stock cut-off, these marks can be made with a rule and chalk or pencil. For greater precision, tools like scribers are used to mark fine lines and hole centers. For still more precision, some industries use lasers and optical instruments.

2. How are layouts classified?

Layouts can be classified by their degree of precision. Precision, as you have seen, is the term used to describe how closely a measurement matches a desired size. **Semi-precision layout** uses measurements with a tolerance of ± 1/64 inch; these can be done with a rule which discriminates to 1/64 in. **Precision layout** uses measuring instruments that discriminate to .001 inch and may have tolerances as fine as .0001 in.

3. What are some tools that are used for layout?

a. Surface plates:

The purpose of the surface plate is to provide **a flat reference surface** from which important dimensions can be measured when laying out a part. These tools are made from granite, cast iron, or steel. Each plate is ground very flat. Plates differ in size from a few square inches, through sizes like 12 by 18 in., up to large sizes like 4 by 12 feet. This important tool is used mainly for precision layout.

surface plate
(SDCCD)

layout dye
(SDCCD)

b. Layout dye:

This liquid, which is most often blue in color, is applied with a brush to the surface of a workpiece. After the dye dries, the machinist lays out the important features of the part on the painted surface, by scratching through the layer of dye to show the metal below; features include hole centers, hole diameters, and lengths and widths of important dimensions.

c. Punches and hammers:

A **pin punch** has a point sharpened to 30°. This punch is used to mark important points along the scribed lines. The pin punch can also mark the centers of holes which are to be drilled or bored into the workpiece.

These same centers are punched next with a **center punch** which has a point sharpened to 90°. The dents will be the starting places and guides for later drilling to make the holes in the workpiece.

The **layout hammer** is used to hit the punches to make the dents.

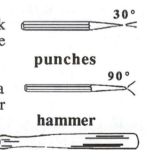

punches

hammer

machinist's scriber
(SDCCD)

d. Scriber:

This is a pointed tool for making lines in the layout dye. It must be sharpened when it starts to get dull, in order to cut thin, clean lines on the surface of the workpiece. Scribers can be tools that are used alone; they can also be included as parts of tools which measure certain dimensions for layout; for example, a vernier height gage has an attached scriber to draw lines.

Directions: Read the following paragraphs and study the pictures:

MORE LAYOUT TOOLS

e. Machinist's combination set:

The machinist's combination set is made up of three different tools mounted on a grooved steel rule. When being used, each tool is mounted separately on the rule, and each serves a different purpose.

machinist's square as a depth gage
(SDCCD)

1. machinist's square: This tool, mounted on the steel rule, is used to measure right angles and 45° angles. It can also be used as a depth gage, as shown. The square contains a **level**, which is a tube of liquid with a bubble in it. When the surface on which it rests is level, the bubble will be centered in the tube.

2. center head: When the center head is used, the edge of the steel rule passes through the center of the head. This allows the machinist to use the center head to find the center of a piece of round stock, simply by drawing two intersecting lines on the end of the workpiece, as shown. That center at the end of the round stock could be center punched and drilled to make center holes. This will be useful for mounting round stock on a lathe.

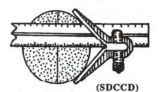

(SDCCD)

center head

3. bevel protractor: When mounted on the steel rule, this tool can be used to lay out or measure angles from 0° to 180° with accuracy to within ± 30'. The picture shows the protractor being used to measure the angle of a bevel.

bevel protractor
(SDCCD)

8-05: "SHOP TALK"   *QUESTIONS and DIRECTIONS*

Directions: Listen to the student's questions and the instructor's directions. Practice pronunciation. Circle the correct directions.

Student Questions: *Instructor Directions:*

1. I'm going to make a gage where the tolerances are ± 1/64 in. Will I need a surface plate for the layout?   A   B   C

2. Next I'll apply layout dye to the workpiece. Any suggestions?   A   B   C

3. I've got the center lines marked on this workpiece. What's next?   A   B   C

4. I want to mark a reference line 1 1/8 inches from the edge of this rectangular part. What tool do I use?   A   B   C

5. What do I do to mark the center of the end of this cylinder?   A   B   C

6. I think this scriber is getting dull. What'll I do?   A   B   C

8-06: EXERCISE

Directions: *Carefully read pages 246 to 250 in the textbook for laying out a gage to measure hole & drill sizes. Your task now is to rearrange and write the eight given layout steps into the correct order. In many operations, you will need to figure out an order for doing things.*

Layout steps for the gage

• Pin punch the hole centers.

• Apply layout dye to the workpiece.

• Measure & scribe the angles & hole centers.

• Cut off a 6 1/16 in. piece of rough stock.

• Study the drawing on page 247.

• Center punch the hole centers.

• Lay down a paper towel.

• Scribe the major reference lines.

(SDCCD)

Steps rearranged in correct order

1. _____

2. _____

3. _____

4. _____

5. _____

6. _____

7. _____

8. _____

8-07: EXERCISE

Directions: *Read pages 246 to 250 again to find the answers to these questions; then write brief answers.*

1. Which tools are used to measure and mark the 1 1/8 width line on the workpiece? _____

2. Which figure best shows that measuring and marking? _____

3. What is the name of the tool that is not one of the tools in the nomenclature section of this lesson? _____

4. What two figures show that tool being used? _____ & _____

5. In the box, make your own drawing of that tool. For help, find a picture of it on page 236.

8-08: EXERCISE

Directions: *Practice spelling some of the nomenclature of layout tools by listening to the spelling and filling in the letters.*

1. p __ n p __ __ c __

2. s __ e __ l __ u __ e

3. l __ __ o __ t d __ e

4. s __ ua __ e h __ __ d

5. __ c __ ib __ r

6. __ e __ e __

7. l __ y __ __ t h __ m __ __ r

8. __ e __ e __ p __ o __ ra __ to __

9. __ om __ in __ tio __ s __ __

10. __ en __ e __ p __ n __ h

11. __ __ r __ a __ e p __ a __ e

12. c __ n __ e __ h __ __ d

89

8-09: VOCABULARY LIST B

Directions: *Study the vocabulary. Write the missing words in the blank spaces.*

1. offset

bent twice to put the scriber point in a different position. *Example*: This scriber is _____ in order to make up for the height added by the base.

2. parallel

running in the same direction and equally apart at every point so as to never meet. *Example*: These precision-ground, flat pairs of rectangular bars are called "parallels," because their sides are _____.

3. accessory

a piece of equipment added to help something work better. *Example*: A right angle plate is not a tool, because it does not measure or shape the metal; but because it supports the work during measuring or shaping, it is called an _____·

(SDCCD)

right angle plate

4. on edge

the position of a part that has been placed on its narrower side. *Example*: Here is a picture of a part that has been placed _____.

5. to elevate

to raise something, to lift something. *Example*: Kim-Sung wants _____ one end of that bar. Let's help her.

6. to stack

to put several things on top of each other in an orderly way. *Example*: Carlos wants _____ several gage blocks to make an elevation of 4.503 inches.

7. to wring

to slide one very clean gage block over another one, so that the two blocks stick together. *Example*: If you are able _____ together the blocks, do not leave them that way when you are finished. Take them apart.

(SDCCD)

Wringing

8. typical

having the qualities of something so as to be a good example of that thing. *Example*: A _____ set of gage blocks has 83 pieces.

9. build-up

several gage blocks stacked on top of each other to reach a desired height. *Example*: Page 173, Figure C-231, in the textbook shows a picture of a gage block _____ used to make a precision height gage.

10. spring-loaded

the quality of any arm or spindle that has a spring behind it that will allow the arm or spindle to return to its original position when pressure is removed. *Example*: The spindle on a dial indicator is _____.

A typical set of gage blocks

90

Directions: *Study the pictures. Listen to the pronunciation of the names of the layout tools and accessories. Practice your pronunciation. Listen a second time and write the words in the blanks.*

NOMENCLATURE OF LAYOUT TOOLS--Part B
(Continued from page 86)

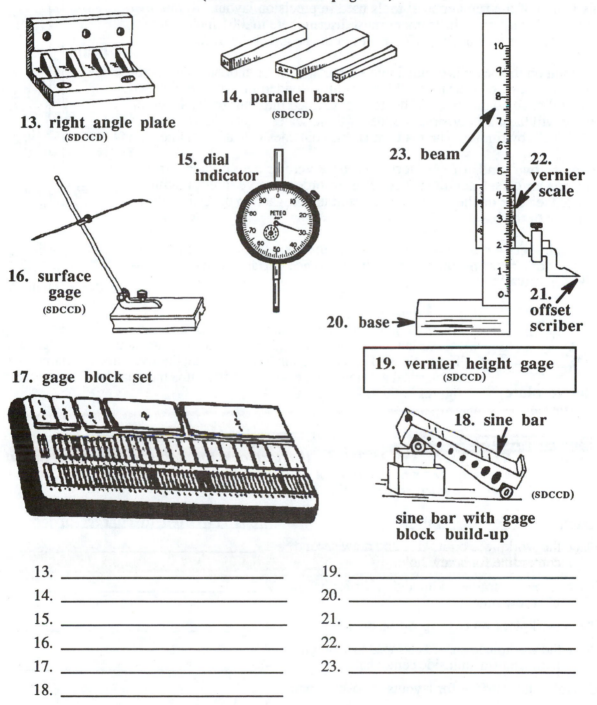

13. **right angle plate**
(SDCCD)

14. **parallel bars**
(SDCCD)

15. **dial indicator**

16. **surface gage**
(SDCCD)

17. **gage block set**

18. **sine bar**
(SDCCD)

sine bar with gage block build-up

19. **vernier height gage**
(SDCCD)

20. **base**

21. **offset scriber**

22. **vernier scale**

23. **beam**

13. _____   19. _____

14. _____   20. _____

15. _____   21. _____

16. _____   22. _____

17. _____   23. _____

18. _____

8-11: READING
Directions: Read the following paragraphs.

MORE LAYOUT TOOLS

4. What are some other tools used in layout?

Vernier Height Gage

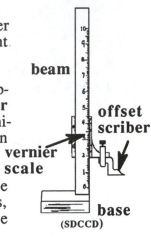
(SDCCD)

a. Vernier height gage:
This is one of the fundamental tools used in precision layout. Remember that a precision layout instrument must discriminate to .001 inch. Height gages range in size from 10 to 72 inches in the height of the **beam**.

Each inch on the beam is divided into ten parts; each tenth of an inch is sub-divided into two or four parts. If the tenth is divided in two, the **vernier scale** will have 50 divisions; if the tenth is divided into four parts, the vernier scale will have 25 divisions. Either of these scales allows discrimination to .001 inch, by finding where a vernier line coincides with a beam line.

There are two kinds of **scribers** used on a vernier height gage. In the picture is an **offset scriber**; it is offset in order to give direct readings, because the scale on the beam does not start until higher up the beam, above the height of the base.

A **straight scriber**, as shown on p. 252 of the textbook, can also be used, but the work must be raised from the reference surface with precision-thickness **parallel bars**.

parallel bars
(SDCCD)

angle plate
(SDCCD)

b. Angle plate:
An angle plate, like the surface plate, is an **accessory.** It is used to help measuring and layout; the angle plate shown in the picture has a right angle. The horizontal arm rests on the surface plate; the workpiece is clamped to the vertical arm to hold the work steady as it is measured and scribed. See figures E-59 to E-62, Text.

8-12: EXERCISE
Directions: Read pages 255 to 258 in the textbook for doing a precision layout of a clamp frame. Your task now is to rearrange and write the five given layout steps in the correct order.

<u>Layout steps for the clamp</u>	<u>Steps rewritten in correct order</u>
• Place the workpiece on its side and draw second center line for screw hole.	1. _____
• Scribe .750 for frame width and 1.625 for height of screw hole.	2. _____
• Place workpiece on edge (position one).	3. _____
• Place the workpiece on end clamped to an angle plate and finish inside center holes.	4. _____
• Use reference surface for layouts in position one.	5. _____

8-13: READING
Directions: *Read the following paragraphs.*

OTHER LAYOUT TOOLS

Surface Gage

c. Surface gage:
The surface gage is pictured in the textbook on page 242, Figure E-24. A scriber can be locked in place at any height along the spindle. It is used to transfer a desired height to the layout by setting the scriber to a rule.

d. Gage blocks:
The purpose of gage blocks is to make international standards of measurement widely available. With gage blocks, individual companies and machinists can compare their working instruments to these blocks to see how close they are to these more exact standards. The gage blocks usually come in **sets** of 81 to 88 blocks; each block differs from the others with thicknesses ranging from .050 to 4.000 inches in a set.

gage block set

Because the gage blocks are so nearly exact in their thickness, they have very small tolerances and can be used to check instruments which discriminate to .001 or even .0001 inch.

In use, the blocks are usually stacked upon each other in combinations that make for a desired height; the process of sliding cleaned blocks together is called **wringing**. If wrung properly, a stack of gage blocks will stick together. How to wring gage blocks is explained on pages 167 to 169 of the textbook.

A list of thicknesses for a typical set of 83 blocks is given on page 171, along with two examples of how the blocks can be combined to make desired thicknesses. In some build-ups, **wear blocks** are placed at both ends of the stack to protect the blocks from wear. Precision height gages can be made from gage block build-ups, as explained on pages 173 and 174 of the textbook.

sine bar

e. Sine bar:
The **sine bar** is a precision-made steel bar that rests on two cylinders, one at each end. The distance between the center lines of the two cylinders is either **5 or 10 inches long**. This length forms the hypotenuse of a right triangle when one end of the bar is elevated while the other end remains touching the surface plate. The elevation is the side opposite the angle formed by the bar and the surface plate; you can call the angle angle A.

As you recall from Unit 6, pages 63 to 66 in this book, the sine function is the ratio of the opposite side to the hypotenuse; therefore **Sine A = opposite ÷ hypotenuse**. If you do a gage block build-up to form the opposite side, you can create an accurate angle for A; the edge of the bar could then be used to scribe that angle on a workpiece. Pages 181 to 183 of the textbook explain.

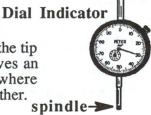

Dial Indicator

f. Dial indicator:
This tool has a spring-loaded spindle that will move in and out when the tip is pressed or released. The movement of the spindle in or out moves an arrow on a graduated dial. This allows for many situations where comparison measurements are possible. Pages 154 to 159 explain further.

Directions: Look at the pictures. Write the name of each tool in the correct lettered space below. Then match letters to the listed purposes.

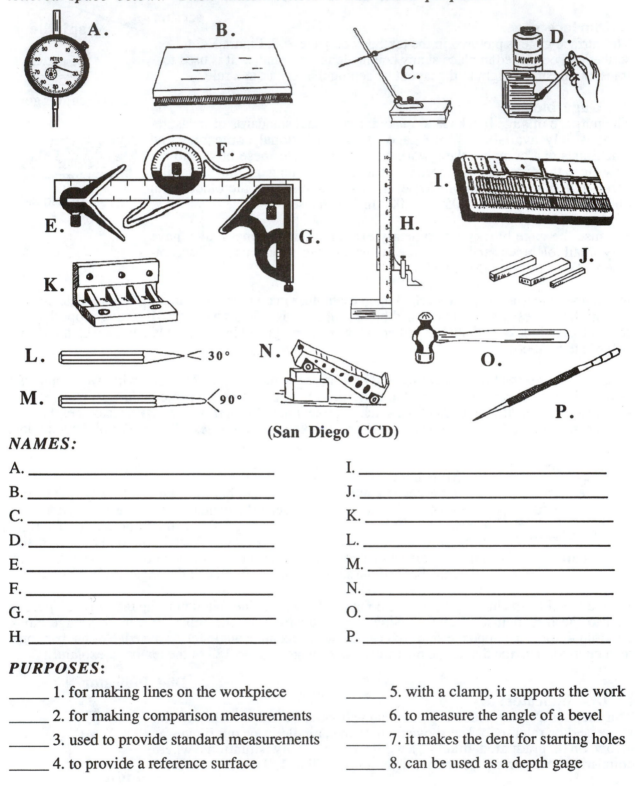

(San Diego CCD)

NAMES:

A. _____

B. _____

C. _____

D. _____

E. _____

F. _____

G. _____

H. _____

I. _____

J. _____

K. _____

L. _____

M. _____

N. _____

O. _____

P. _____

PURPOSES:

_____ 1. for making lines on the workpiece

_____ 2. for making comparison measurements

_____ 3. used to provide standard measurements

_____ 4. to provide a reference surface

_____ 5. with a clamp, it supports the work

_____ 6. to measure the angle of a bevel

_____ 7. it makes the dent for starting holes

_____ 8. can be used as a depth gage

94

Unit 9: *SAWING MACHINES*

ASSIGNMENT: **Read, study and complete pages 95 to 102 of this book. Then read pages 302 to 344 in the textbook, *Machine Tool Practices, SECTION G, Introduction, and Units 1 to 4.***

OBJECTIVES for this unit:
You should be able to:
1. Given pictures and illustrations, identify several different types of saws and their uses.
2. Recognize and correctly use some sawing terms found in the textbook *Glossary*.
3. Learn the importance of study outlines and complete an outline from the textbook's Section G.

9-01: VOCABULARY LIST A
Directions: Study the vocabulary.   Write the missing words in the blank spaces.

1. to cut off

to cut through rough stock to get a piece of usable length. *Example:* A horizontal band saw can be used _____ a piece of rough stock.

2. contour

the outline or shape of something; to make something so it has a particular shape.   *Example:* Naomi used the saw to cut the basic _____ of the clamp frame.

driven wheel

3. intricate

containing many unusual twists and turns of shape. *Example:* Bill used the vertical band saw to cut the _____ contours of the workpiece.

band saw blade

4. band saw blade

a continous loop of steel with teeth on one edge; a wheel is driven to move the blade in a continuous horizontal or vertical path for cutting. *Example:* Le measured the _____; then he welded it into a loop.

5. reciprocating

moving alternately back and forth. *Example:* The motion of your hand using a hacksaw is a _____ motion.

6. hinge

a device that joins two parts, but leaves them free to turn or swing. *Example:* Some cuttoff saws have a _____ which allows a reciprocating saw blade to be lowered for a cut.

7. to tilt

to move something so that it slopes down. *Example:* Some saws will _____ in order to cut the stock at an angle.

8. abrasive

a substance such as finely ground aluminum oxide used for grinding, smoothing, or polishing. *Example:* Each little piece of _____ has sharp edges which act as cutting teeth for removing material.

9. hydraulic

moved by the pressure of a liquid forced through an opening or tube. *Example:* Some saws move to make cuts. Some _____ worktables move the work through the path of the saw blade.

10. velocity

speed. *Example:* Cutting at 15,000 feet per second is high _____.

Directions: Read the following paragraphs.

AN IMPORTANT STUDY TOOL: THE WRITTEN OUTLINE

1. What is an outline?

An outline is a summary of a subject made up of a systematic listing of its most important points.

a) An outline is a **summary**; you make judgements about what is most important in a longer piece of writing and write down those points in shorter form.

b) An outline is a summary of **one subject**; everything in the outline must be related to the subject.

c) An outline lists the important points in a **systematic** way; the system uses letters and numbers to show how some ideas are equal in importance and how some others are subdivisions of larger ideas and points. Here is a numbering and lettering system often used in outlines:

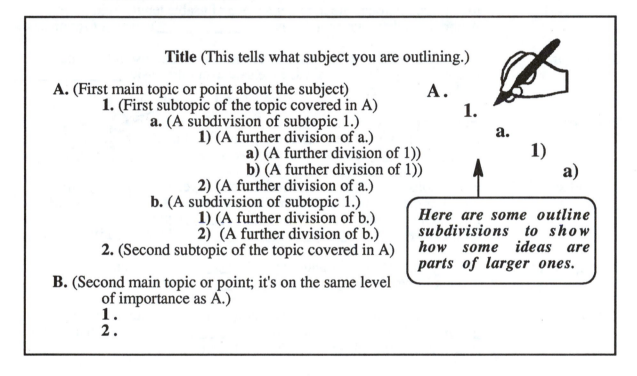

Title (This tells what subject you are outlining.)

A. (First main topic or point about the subject)
    1. (First subtopic of the topic covered in A)
        a. (A subdivision of subtopic 1.)
            1) (A further division of a.)
                a) (A further division of 1))
                b) (A further division of 1))
            2) (A further division of a.)
        b. (A subdivision of subtopic 1.)
            1) (A further division of b.)
            2) (A further division of b.)
    2. (Second subtopic of the topic covered in A)

B. (Second main topic or point; it's on the same level of importance as A.)
    1.
    2.

Here are some outline subdivisions to show how some ideas are parts of larger ones.

2. What basis is used to divide a topic into smaller parts?

It is possible to divide a topic several ways using a different basis for each way. For example, if you were dividing the subject "tools" into subtopics, you could use different bases for dividing.

Topic	Subtopics	Basis of division
TOOLS	1. expensive tools 2. inexpensive tools	the cost of tools
TOOLS	1. hand tools 2. electric-powered tools	the power source of tools

9-02: READING (continued)
Directions: Read the following paragraphs.

3. Is there a general pattern of how the textbook is organized and subdivided?

You will see the pattern of how your textbook is organized by looking at the *CONTENTS* pages at the front of your text. There are fifteen major *SECTIONS* (*A* through *O*). Each section title has its own capital letter name and is printed in red/brown capital letters to show that the sections are on the same level of importance. The units are the next subdivision of the sections. As you look through the contents and the text of the book, you will find that most sections will give you subdivisions like these:

- Introduction to a general class of tools with the general purpose given.
- Particular classes of tools that come under the general class.
- A list of the important parts of a larger tool (for example, motor, cutting mechanism, and feed mechanism) or the members of a class of tools (for example, files, pliers, screwdrivers, hammers, and punches).
- The purpose of each part of a larger tool or the purpose of some smaller individual tool.
- The major accessories used with the tools.
- Typical operations in which the tool can be used.
- Safety precautions for using a particular tool.

4. Why is making an outline of the sections and units a help to studying?

Making a written outline can help you in several ways:

1. You learn how the authors have organized the material and what they think is most important. You have to make judgements about the importance of items in the text.
2. You have to understand the major ideas and put them in your own words or find the key words of the authors.
3. Then you will have, in shorter form, a record of the most important ideas.
4. The outline becomes a good tool for study, review, and preparation for tests.

9-03: EXERCISE
Directions: Complete the outline below with information from the textbook pages to which you are referred. The basis for outlining the topic is "types of saws."

SECTION TITLE: _G_____ (page 302)

A. Classes of sawing machines:

    1."The first class is _____." (page 302, paragraph 1)
       Purpose: "is to reduce mill lengths of _____ _____ material into lengths
       suitable for holding in other machine _____." (p. 302, paragraph 2)

    2. "The second class is the _____. (p. 302, paragraph 1)
       Purposes:
       a. "A piece of stock material can often be _____ to final _____ by one or
          two saw _____. (p. 308, paragraph 2)

97

9-03: EXERCISE *(continued)*
<u>*Directions*</u>: *Continue making the outline, as on the previous page.*

        b. "A second important advantage...is **contouring ability**..., the ability to cut
            intricate _____ shapes." (p. 308, paragraph 3)

B. Types of Cutoff Machines:
    1. _____ *SAW   or power hacksaw* . (page 302, paragraph 3)
        a. _____*hinge type*_____
        b. _____ (p. 302, paragraph 4)

    2. _____*Band*_____ (p. 303)
      Note: This machine "uses a steel band _____ with the _____ on
      one _____." (p. 303, paragraph 1)

    3. _____*Frame*_____ (p. 304, paragraph 2)

    4. _____ (p. 304, paragraph 3)
      Note: This machine "can be used to cut a number of _____
      materials such as _____ , _____ , and _____."

    5. _____ (p. 304, paragraph 4)
      Notes: This machine "uses a _____ metal saw with teeth."

C. Types of Vertical Band Machines (p. 309)

    1. General-Purpose _____ (p. 309, paragraph 2)
      Notes: a. "The general-purpose band machine is found in most machine shops."
            b. This machine has "a worktable that can be _____ in order
              to make angle cuts. (p. 311, paragraph 1)

    2. Band Machines with Power-Fed Worktables   (p. 311, paragraph 3)
      Note: "The _____ is moved hydraulically. The _____ is
      relieved of the need to _____ the workpiece into the cutting blade.

    3. _____ -Tool- _____ Band Machines   (p. 311, paragraph 4)
      Note: "band speeds can range as high as _____ to _____ feet per minute.

    4. Large-_____ Band Machines (p. 312, paragraph 2)
      Notes: "This type...is used on _____ workpieces...The workpiece
        remains _____ and the saw is _____ about."

9-04: QUIZ
<u>*Directions*</u>:   *For each statement, circle T for "True" or F for "False."*

T F 1. The two general classes of saws are *cutoff saws* and *vertical band saws*.
T F 2. The reciprocating saw is the most common cutoff tool in the machine shop.
T F 3. Vertical band saws are often used to cut out the basic shape of a part.
T F 4. A band saw uses a continuous steel band blade with teeth on one edge.
T F 5. Horizontal band saws are used mostly for cutoff of correct lengths from bar stock.
T F 6. Abrasive saws are the most common saw in the machine shop.
T F 7. Some of the saws in the outline will be used more frequently than others.

9-05: VOCABULARY LIST B

Directions: *Find each vocabulary word from the "Glossary" in the text-book, pages 801 to 808; then fill in the blanks. Check your answers at the back of this book.*

1. cutting fluid   any of several materials used in _____ metals: cutting _____, synthetics, _____ or emusified oils (water based) and sulfurized oils. *Example*: The machinist sprayed the _____ on the blade of the saw.

2. coolant   a cutting _____ used to cool the tool and the workpiece, especially in grinding operations; usually _____. *Example*: The _____ is sprayed on the the workpiece through a nozzle.

3. roughing   in machining operations, the rapid _____ of unwanted material on a _____, leaving a _____ amount for _____. *Example*: A band saw can be used for _____ out the basic shape of the workpiece; it can then be finished on other machines.

4. kerf   the _____ of a _____ produced by a saw. *Example*: Page 320, Figure G-48, in the textbook shows a comparison of _____ widths; the _____ of an abrasive saw is wider than the _____ of a band saw.

kerf →

5. gullet   the _____ of the space between _____ on saws and circular milling cutters. *Example*: Both the saw blade and the milling cutter have a _____.

6. rake   a tool _____ that provides a _____ (sharpness) to the cutting edge. *Example*: The picture shows: (1) positive _____, (2) zero _____, and (3) negative rake. Positive rake cuts faster.

7. relief angle   an _____ that provides cutting edge _____ for the cutting action. *Example*: Because it provides a space for material to go after it is cut, angle **R** is called a _____ or a clearance angle.

rake angle

relief angle or clearance angle

R

gullet

positive rake   negative rake
zero rake

SOME TERMS FROM SAWING

8. pitch & width   in saw _____ , the number per _____. *Example*: If there are eight teeth in one inch on a saw blade, you have a _____ of 8, and each tooth is 1/8 inch apart. The _____ is from the flat edge to the tip of the tooth on the sawing edge.

pitch:
number of teeth per inch

width

1 inch

9. set
   &
   gage

the width of saw teeth; it is measured
from the tip of a tooth on one side
to the tip of a tooth on the other side.
The set of saw teeth is _____
than the gage width. *Example*:
To learn some _____ patterns,
look at Figure G-41 on text page 317;
the _____ is the thickness of the blade; it is not the same as the width.

set dimension   gage

10. to anneal   a heat treatment in which _____ are heated and then _____ very
slowly for the purpose of decreasing _____. *Example*: After a bandsaw blade has been welded, the weld is very hard and not easy to bend. Therefore,
we will have _____ it, so it will not break when put in the saw.

9-06: QUIZ

Directions: *For each statement, circle T for "True" or F for "False."*

T F 1. The set of saw teeth is wider than the gage.
T F 2. The set dimension is the total distance from the tip of a tooth on one side to the tip of a
         tooth on the other side.
T F 3. The kerf of a band saw is usually wider than the kerf of an abrasive saw.
T F 4. A saw with a blade pitch of 4 has teeth that are one quarter inch apart.
T F 5. After a bandsaw blade has been welded, it must be annealed.
T F 6. Finishing cuts with machine tools usually come after roughing with a saw.
T F 7. A positive rake on a saw tooth does not cut any faster than a negative rake.

9-07: NOMENCLATURE EXERCISE

*Directions: Study these pictures as you listen to the pronunciation
of these saws and sawing terms. Listen again and write the terms.*

A. hinge-type reciprocating
   cutoff saw

D. coolant
   nozzle

B. horizontal band
   cutoff saw
   (SDCCD)

C. general purpose
   vertical bandsaw

D. coolant
   nozzle

A. _____

B. _____

C. _____

D. _____

9-07: NOMENCLATURE EXERCISE (continued)

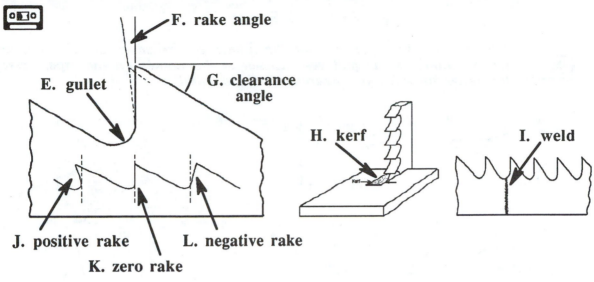

E. gullet   F. rake angle   G. clearance angle   H. kerf   I. weld   J. positive rake   K. zero rake   L. negative rake

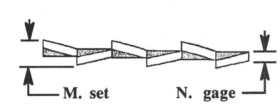

M. set   N. gage

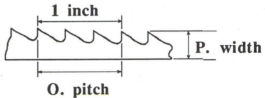

1 inch   P. width   O. pitch

E. _____

F. _____

G. _____

H. _____

I. _____

J. _____

K. _____

L. _____

M. _____

N. _____

O. _____

P. _____

9-08: CONVERSATION

Directions:
Listen to Lucy Garcia and Le Tran talking about saws. Fill in the missing words. Rewind and listen again. Practice pronunciation.

Lucy: Le, I saw you using a reciprocating saw for your _____.

Le: Yes, I like the way it cuts through the _____.

Lucy: I'm learning how to use a _____ bandsaw for cutting contours. I've been using it to _____ _____ some workpieces. I use an 8 or a 12 _____ blade.

Le: So do I. I use the _____ pitch because it cuts through the _____ material more easily.

Saws and Sawing

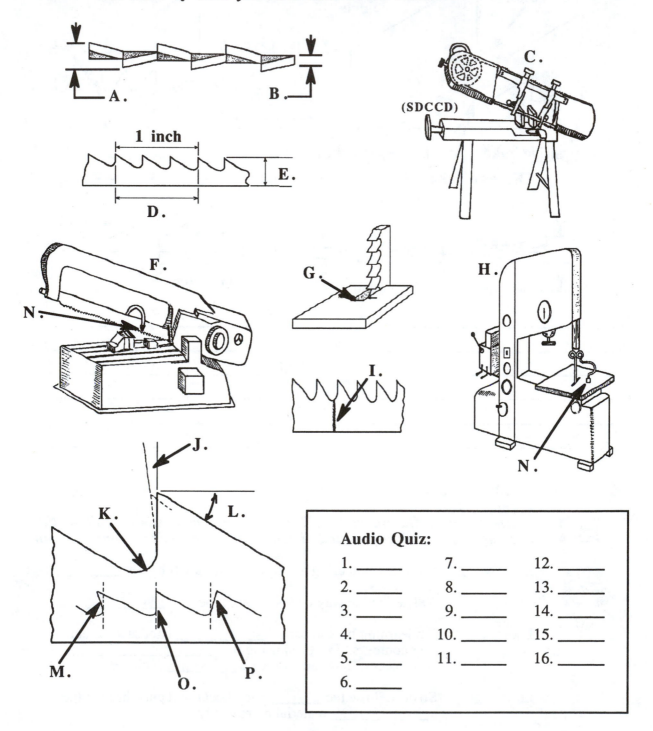

Directions: *Test yourself on the names of the saws and the sawing terms. Look at the pictures. Listen to the names on the tape. Write the letter of what you hear next to the number.*

Audio Quiz:

1. _____   7. _____   12. _____
2. _____   8. _____   13. _____
3. _____   9. _____   14. _____
4. _____   10. _____   15. _____
5. _____   11. _____   16. _____
6. _____

Unit 10: *DRILLING MACHINES and DRILLING*

ASSIGNMENT: **Read, study and complete pages 103 to 118 of this book. Then read pages 345 to 388 in the textbook,** *Machine Tool Practices,* *SECTION H, Drilling Machines.*

OBJECTIVES for this unit:
You should be able to:
1. Identify and correctly say the important parts of the sensitive drill press.
2. Identify the names of some major drilling operations and the tools used.
3. Identify and correctly say the major parts of the twist drill.

10-01:   VOCABULARY LIST   A
Directions: Study the vocabulary.   Write the missing words in the blank spaces.

1. press   any machine by which something is squeezed, stamped, smoothed, fitted, or drilled by pressure. *Example:*
The drill _____ uses pressure to send a rotating twist drill into a workpiece in order to make a hole.

straight shank

twist drill

2. capacity   the amount that can be received or held. *Example:* Al told the students that the size of drill presses is measured by their _____.

3. base   the part on which something rests. *Example:* The _____ of a drill press is made of heavy cast iron, so the machine will stand straight.

4. column   a slender, upright, cylindrical shaft which is strong enough to support the heavy objects that are attached to it. *Example:* In a drill press, the base supports the _____, which supports the table and the drill head.

5. chuck   a metal clamping device used to hold a rotating tool (like the twist drill), or rotating work (like a round workpiece on a lathe). *Example:* Igor put the twist tdrill into the _____ and tightened it with a _____ key.

chuck key

drill chuck

6. sensitive   being aware of how something feels.   *Example:* In this unit, you will study the _____ drill.

7. pulley   a small wheel with a grooved rim in which a belt runs. *Example:* In the sensitive drill, the motor turns a vee _____ which turns a vee belt which turns another vee pulley.

pulleys

8. spindle   in a drill, the spindle is a rotating tube into which tapered chucks or tapered drills can be inserted. *Example:* The chuck which holds a straight twist drill has a tapered end which fits into the taper of the _____.

9. controls   instruments or devices used to direct the operation of a machine. _Example_:
                  The major machines usually have two sets of _____: one set
                  is manual; the other set is automatic.

10. automatic   done by the power of a machine motor; not done by hand. _Example_: An
                  _____ control uses electrical power; a manual control
                  needs the hands of the operator to do turn a handle or push a lever or do
                  similar activities.

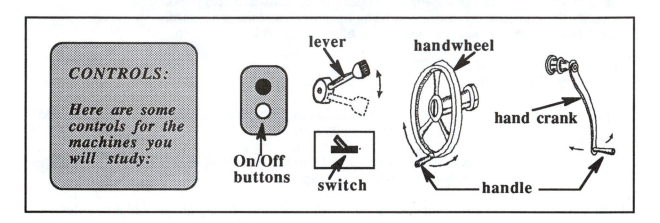

10-02: "SHOP TALK"   *QUESTIONS   and ANSWERS*

Directions: **Listen to the students' questions and the instructor's answers on the tape. Practice your pronunciation as you repeat each question and answer. Fill in the blanks with what you hear.**

Student: How do you turn on this _____ drill?

Instructor: Look up toward the top. There's a _____ and a _____ button. Push
                the green button to _____ the machine and the red button to _____ it.

Student: Hey, that's easy. Can you tell me about some of the other _____?

Instructor: Yes, this drilling machine and the other larger machines all have controls for
                directing the machine what to do. There's usually _____ sets of controls.

Student: I was reading about that this morning. There's usually a set of _____
                controls that use electrical power and another set of _____ controls that
                require the operator to turn or pull something.

Instructor: Yes! Very good! Those are the two big categories of controls: automatic and
                manual. For right now we'll stick with the _____ ones. For example,
                those three levers on the right side of the machine are the manual _____; that's
                where the "sensitive" part of sensitive drill comes in. You will learn to get a "feel"
                for how much _____ to put on those levers to get the right feed.

Student: Thank you very much, teacher. I'm going to learn the rest of the names for these
                _____ _____.

Directions: *Read the following paragraphs.*

BASIC TYPES OF DRILL PRESS

The **purpose** of any **drill press** is to drill holes in workpieces. After a hole is drilled, other operations may be done in the hole; for example, the machinist may thread the inside of the hole, or ream it to get a more exact, smoother inside diameter.

The **capacity** of any drilling machine is measured by the largest diameter of a circular workpiece that can be drilled in the center; that is the same as measuring from the center of the drill spindle to the column. Capacity is a way to talk about the size of a drilling machine.

There are three basic types of drill press; look at the first one:

1. The SENSITIVE DRILL:

This drilling machine is called "sensitive," because it has a hand lever on the side which allows the machinist to feed the drill into the work by hand. With practice, you, or any other machinist, will get a "feel" for how much pressure to apply.

Some of the main parts of the sensitive drill include: the **base** and the **column,** which support the weight of the other parts of the machine, and the **table,** to which the work is secured. Another main part is the **drilling head** which houses a system of **pulleys** and **vee-belts** which drive the **spindle**; the spindle is a tube which has a standard Morse taper (5/8 inch of taper per foot) on the inside. The spindle holds a **drill chuck** which has a shank with a **Morse taper**; when the chuck shank is pushed into the spindle the tapers will hold the two pieces together. The chuck has three jaws into which a **twist drill** with a straight shank is pushed; the jaws are then tightened with a **chuck key** to hold the drill. If the drill has a tapered shaft it can fit directly into the spindle, without the chuck. The twist drill cuts the hole into the workpiece.

Many drilling machines have a **depth stop** which regulates how far a drill can go down into the work. An electric **motor** drives the vee belts and pulleys. Some drilling machines use a motor, and a **variable speed drive** instead of the vee pulley and the vee belts.

Various **controls** are located on the machine to control the drilling: You should know where the **ON/OFF buttons** are located; the drill can be fed into the work by the hand lever or by an automatic **power feed.** A **speed control** is necessary to control how fast the spindle is turning.

The sensitive drill can stand on the floor or a work bench.

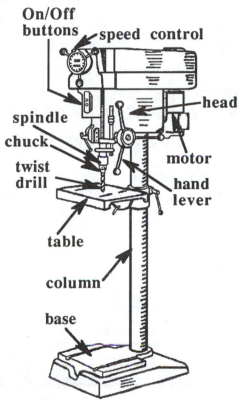

Sensitive Drill Press
(San Diego CCD)

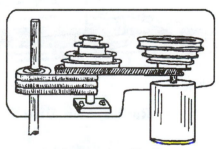

vee-belts, pulleys, and motor

105

10-04: EXERCISE
Directions: *Read the questions. Circle "T" for TRUE or "F" for FALSE.*

T F   1. The capacity of a drilling machine measures how many parts per minute the machine will make.

T F   2. After a hole is drilled, other operations may be done in the hole.

T F   3. To operate a sensitive drill successfully usually requires enough practice to get a "feel" for how much feed to give using the hand levers.

T F   4. Sensitive drills are powered by vee pulleys and belts or by a variable-speed drive.

T F   5. The work should be secured to the table before the sensitive drill is used.

T F   6. A twist drill with a straight shank is held in place by a drill chuck.

T F   7. The chuck key loosens and tightens the three jaws of the drill chuck.

10-05: READING
Directions: *Read the following paragraphs.*

2. The UPRIGHT DRILL PRESS:

This drill press is the second kind of drill press you are likely to see in the shop. It is very similar to the sensitive drill press in its major parts, but it is bigger, more powerful, and capable of doing heavier work than the smaller drill; for example, it might be used to drill hole diameters to 2 inches or greater. For a picture (Fig. H-16) and further details, see the textbook, pages 350 to 353.

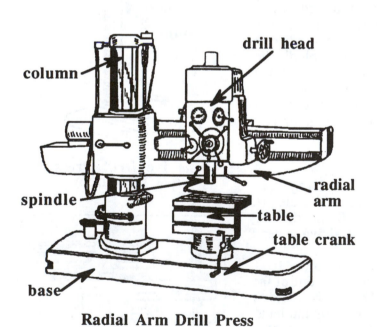

Radial Arm Drill Press

3. The RADIAL ARM DRILL PRESS:

The radial arm drill press is a large drilling machine with a capacity for making holes in much larger work than the first two machines; that is because of the **radial arm** which puts the **spindle** at a much further distance from the **column**.

The operator can also secure a heavy workpiece to the table. Then he/she can position the radial arm over the work and slide the **drill head** along the arm to a position for drilling. The radial arm will also slide up and down on the column.

The radial arm drill press has a variety of controls, like the **table crank** that raises and lowers the table.

10-06: EXERCISE

Directions: _Read the questions. Circle "T" for TRUE or "F" for FALSE._

T   F   1. The upright drill press is for doing work that is too heavy for the radial arm drill press.

T   F   2. The radial arm drill press has a larger capacity than the sensitive or upright drill presses.

T   F   3. One disadvantage of the radial arm drill is how difficult it is to reach the work.

T   F   4. The radial arm drill press is different from the other two, because they do not have radial arms.

T   F   5. An upright drill press can drill holes with a diameter of two inches.

10-07: MACHINE NOMENCLATURE

Directions: _Study these pictures as you listen to the pronunciation of these parts and controls. Rewind and practice your pronunciation._

Drilling Machine Main Parts

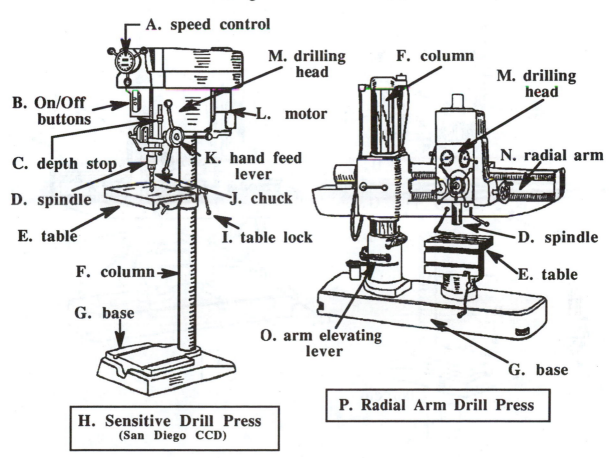

A. speed control
M. drilling head
F. column
M. drilling head
B. On/Off buttons
L. motor
C. depth stop
K. hand feed lever
N. radial arm
D. spindle
J. chuck
E. table
I. table lock
D. spindle
F. column
E. table
G. base
O. arm elevating lever
G. base

P. Radial Arm Drill Press

H. Sensitive Drill Press
(San Diego CCD)

10-07: SPELLING

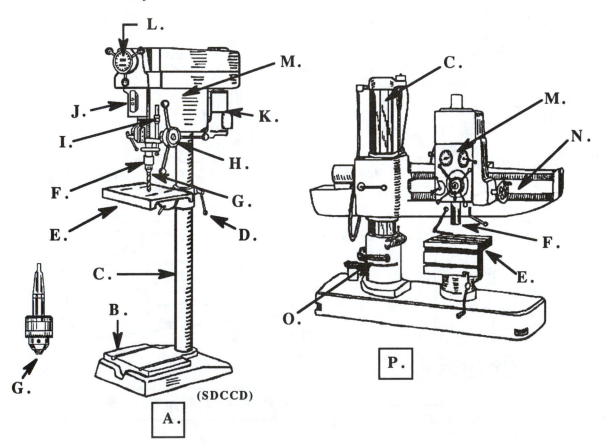

Directions: **Rewind, and listen to the words again. Fill in the missing letters.**

A. s _ ee _   c _ n _ _ ol

B. O _ /O _ _   b _ t _ o _ s

C. _ e _ t _   s _ o _

D. s _ in _ l _

E. _ a _ _ e

F. _ o _ u _ _

G. _ a _ e

H. s _ n _ i _ e   _ r _ ll   p _ e _ s

I. t _ _ l _   lo _ _

J. _ h _ _ k

K. h _ n _   _ ee _   le _ e _

L. _ o _ o _

M. d _ i _ lin _   h _ _ d

N. r _ d _ _ l a _ _

O. _ r _ e _ e _ a _ i _ g   _ e _ er

P. r _ d _ a _   _ r _   _ r _ l _   p _ es _

10-08: MACHINE NOMENCLATURE MATCHING EXERCISE

Directions: **Continue learning the names of the drill press parts and controls. Look at the pictures. At the top of the next page, write the correct letter by each name.**

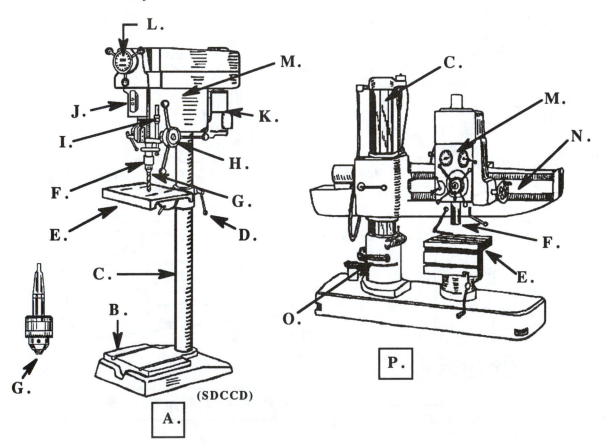

(SDCCD)

MATCHING: *Write the letters from the pictures next to the names.*

____ 1. spindle   ____ 7. radial arm   ____ 12. radial arm drill press

____ 2. chuck   ____ 8. table   ____ 13. speed control

____ 3. base   ____ 9. drilling head   ____ 14. sensitive drill press

____ 4. column   ____ 10. On/Off buttons   ____ 15. hand feed lever

____ 5. depth stop   ____ 11. table lock   ____ 16. arm elevating lever

____ 6. motor

10-09: VOCABULARY LIST B
Directions: *Study the vocabulary.   Write the missing words in the blank spaces.*

1. to mate — to fit one part into another that is shaped to receive it. *Example*: The shank of the chuck is tapered _____ with the taper of the spindle.

2. to deburr — to use a file or other tool to remove sharp, metal burrs from a workpiece. *Example*: Al showed me how _____ the newly drilled holes.

3. to guide — to lead in a particular direction. *Example*: The handrail is placed near the stairs _____ your hand and you to the bottom of the stairs.

4. to align — to bring a variety of parts into a straight line. *Example*: The machinist has _____ the chuck, the drill, and the hole center in the work.

5. cone — a solid figure from geometry, having a circular base and sides that taper evenly to a point; anything shaped like that figure. *Example*: The countersink drilling tool cuts a _____ shape at the top of a previously drilled hole.

6. pilot — a part of a cutting tool that guides the tool into a previously drilled hole. *Example*: The end of the drill chuck key is a _____ which guides the gear teeth of the key into the teeth of the chuck.

7. tip — the pointed, tapering, or rounded end of something long and slim. *Example*: There is a chamfer at the _____ of the reamer.

8. clearance — the clear space between moving objects or parts. *Example*: In machining, all cutting operations will require _____ behind the cutting edge to allow the cut material to pass by.

9. adjustment — making a machine accurate by changing the controls. *Example*: Sometimes the machinist must make an _____ in the drill's speed.

10. axis — an invisible line that passes through the center of a ball, cylinder, or taper. *Example*: A twist drill has an _____ that runs through its center. (The plural of *axis* is *axes*.)

109

10-10: READING
Directions: *Read the following paragraphs.*

DRILL PRESS OPERATIONS

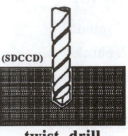

twist drill

1. For drilling holes, use a twist drill:
Twist drills come in a variety of sizes so the machinist can find the right drill for the diameter dimensions shown on engineering drawings. Twist drill shanks are straight or have a taper which will mate with the taper of a spindle.

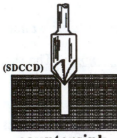

countersink

2. For countersinking, deburring or chamfering, use a countersink:
A countersink is used to make a cone-shaped cut at the top of a hole which has been drilled previously. Why do this? Reasons are:
(1) to chamfer the existing hole (a chamfer is most often 1/32 to 1/16 inch wide), (b) to deburr a hole, or (3) to countersink a hole deep enough to receive the head of a flathead screw. The angle on a countersink point can be 60, 82, 90, or 100 degrees; for example, to make a 45° chamfer, you would used a 90-degree countersink.

3. For counterboring holes, use a counterbore:
This tool is used to enlarge an existing hole. The counterbore has a pilot as a tip; the pilot guides the cutting part of the counterbore down into the hole, with correct centering. The resulting enlargement of the hole has a flat bottom. The counterbore can also be used to cut a recess for a bolt head and certain screws which need to lie below the surface of the work. The pilot will need .002 to .005 in. of clearance in the previously drilled hole.

counterbore

4. For starting a tapping operation, use the drill press to align the hand tap:
When tapping interior threads in a previously-drilled hole, the drill press can be used to correctly align the first tap: the tap, with a tap handle, is placed in the chuck, and the spindle is turned by hand rather than by starting the motor. After the tap is started, the tapping procedure is completed by hand. See Figure H-101 in the textbook.

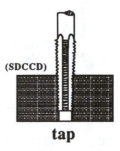

tap

reamer

5. For making smooth, accurately-located holes, use a reamer:
The purpose of a reamer is to enlarge previously-drilled holes or give them a precision finish; the resulting holes will be accurate, straight, and uniform in diameter. Many reamers have a 45° chamfer at their tips which helps in the cutting. Machinists must make adjustments in the speed and feed when using a reamer: the speed will be slower and the feed will be greater than for drilling the original hole. Using good cutting fluids is very necessary. There are many different kinds of reamers shown on pages 385 and 386.

110

Directions: **Read the questions. Listen to three possible answers. Circle the best.**

1. What is the name of the part on a counterbore which guides it into the hole?   A   B   C

2. Which operation of a drill press is used to make a seat for the head of a flathead screw?   A   B   C

3. Four of these operations require a previously-drilled hole. Which tool is used to make those first holes?   A   B   C

4. Why is it good to start a hand tap in a drill press?   A   B   C

5. What is the purpose of tapping an existing hole?   A   B   C

6. What operation will make a recess for a bolt head?   A   B   C

7. Which one of the procedures above can give precision results?   A   B   C

10-12: VOCABULARY LIST C
Directions: **Study the vocabulary. Write the missing words in the blank spaces.**

1. edge
the thin cutting part of a blade. *Example*: The _____ of the butcher's knife was worn, so he had to sharpen it for cutting the meat.

2. to sharpen
to grind the edge of a cutting tool to make it thin and better able to cut. *Example*: You need _____ the cutting edges of this drill.

3. chisel
a sharp-edged tool for cutting or shaping wood, stone or metal. *Example*: Michaelangelo used a _____ to make beautiful statues.

4. to slope
to be or run in a direction that is not flat or level. *Example*: After you pass the ridge, the road begins _____ down into the valley.

5. helix
a continuous line that goes around and down a cylinder or cone, like a screw thread. *Example*: The flutes on a twist drill are cut in the shape of a _____. An adjective from this word is *helical*.

6. cutting lips
the sharp edges at the top of a twist drill which continue the cutting action begun by the chisel edge. *Example*: In order to get a smooth cut, the _____ of a twist drill should all be of the same length.

7. relief
clearance, the space behind a cutting edge, to allow the cut material to slide by. *Example*: The _____ angle is the same as the clearance angle.

111

10-13:   TWIST DRILL NOMENCLATURE

The Parts of the Twist Drill--Part I

One of the skills you will want to develop is sharpening your own twist drills. This task will be easier if you know the nomenclature for the twist drill. When your teacher shows you how to do this or comments on the job you have done, you will understand the words.

Directions: Study these pictures as you listen to the pronunciation of these parts of the twist drill. Pronounce each part's name after you hear it. Rewind, listen again, and complete the exercises below.

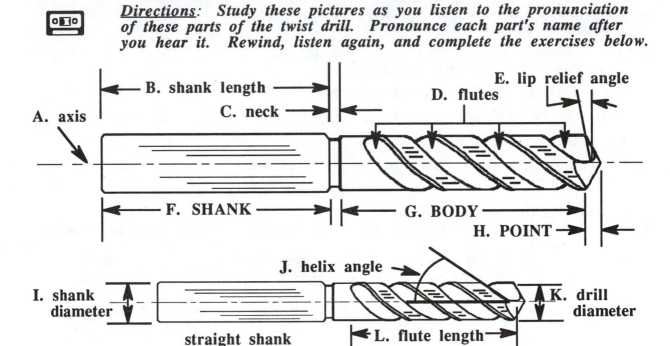

SPELLING: *Directions: Listen to the words again. Fill in the missing letters.*

A. a __ i __

B. s __ an __   l __ n __ t __

C. __ e __ k

D. f __ __ u __ e __

E. l __ p r __ l __ e __   __ n __ __ e

F. __ H __ N __

G. __ O __ Y

H. __ O __ N __

I. __ __ a __ k d __ a __ e __ e __

J. h __ li __   a __ g __ e

K. __ r __ l __   __ i __ m __ t __ r

L. __ l __ t __   __ en __ t __

WRITING: *Directions: Listen to the words again. Write the words.*

A. _____
B. _____
C. _____
D. _____
E. _____
F. _____
G. _____
H. _____
I. _____
J. _____
K. _____
L. _____

112

TWIST DRILL NOMENCLATURE EXERCISE

The Parts of the Twist Drill--Part I

MATCHING
Directions: *Look at these pictures. Write the correct letter next to each name below.*

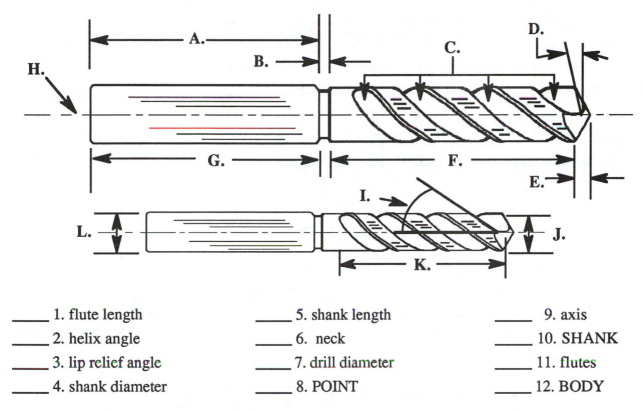

_____ 1. flute length	_____ 5. shank length	_____ 9. axis
_____ 2. helix angle	_____ 6. neck	_____ 10. SHANK
_____ 3. lip relief angle	_____ 7. drill diameter	_____ 11. flutes
_____ 4. shank diameter	_____ 8. POINT	_____ 12. BODY

10-15:   READING
Directions: *Read the following paragraphs.*

The purpose of the twist drill is to drill holes into workpieces. The three main parts of the twist drill are: (1) the **shank**, (2) the **body**, and (3) the **point**. The body is separated from the shank by a small recessed part called the **neck**.

An axis line runs the length of the drill through its center; the **axis** is important, because all **parts** of the twist drill are related to that center; if those drill parts are not aligned, the machinist may produce holes that are "out of round," that is, not perfectly round, or not in the right position.

Cut into the sides of the drill body are one or more helix-shaped **flutes** (the picture shows a drill with two.) The flutes have the double purpose of: (1) allowing the material which is being cut by the **point** to be pushed out of the hole in the form of a chip curl; and (2) allowing cutting fluid to flow down the flutes to aid in smooth cutting.

The shape of the flute's helix is measured by the **helix angle**. When you study the point, you will see more about the lip relief angle.

10-16: **EXERCISE**

Directions: Fill in the blanks with words from the word list.

> **Word List:** chips, axis, cutting fluid, two, neck, helix angle, one, hole, point, helix angle, shank, center, flutes, body

The three main parts of a twist drill are: (1) the _____, (2) the _____, and (3) the _____. A line that runs through the center of the twist drill is called the _____. In the picture, this line is shown by long lines alternating with dashes; that is a _____ line. Cut into the sides of the twist drill body are the _____. There can be _____ or more of these. They have two important purposes: (1) they allow the curling _____ to be pushed out of the _____; and (2) they allow _____ to flow to the place where the cutting is taking place. The shape of the flute's helix is measured by the _____ _____.

10-17: **TWIST DRILL NOMENCLATURE**

The Parts of the Twist Drill--Part II

Directions: Study these pictures as you listen to the pronunciation of these parts of the twist drill. Pronounce each part's name after you hear it. Rewind, listen again, and complete the exercises on the next page.

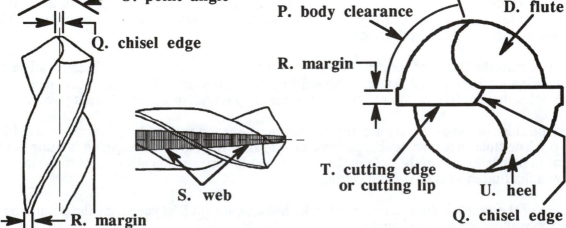

114

SPELLING: *Directions:* Listen to the words again. Fill in the missing letters.

E. l__p r__l__e__ c__ea__an__e   R. __a__g__n

M. t__n__   S. __e__

N. __a__e__ s__ __n__   T. c__t__ __n__ e__ __e

O. __oi__t __n__l__   or cu__ __in__ l__ __

P. b__d__ c__ __a__a__ __e   U. __ee__

Q. c__i__e__ e__ __e   D. f__ __t__

WRITING: *Directions:* Listen to the words again. Write the words.

E. _____   R. _____

M. _____   S. _____

N. _____   T. _____

O. _____   or _____

P. _____   U. _____

Q. _____   D. _____

10-18: TWIST DRILL NOMENCLATURE EXERCISE

The Parts of the Twist Drill--Part II

Directions: Look at these pictures. Write the correct letter next to each name below.

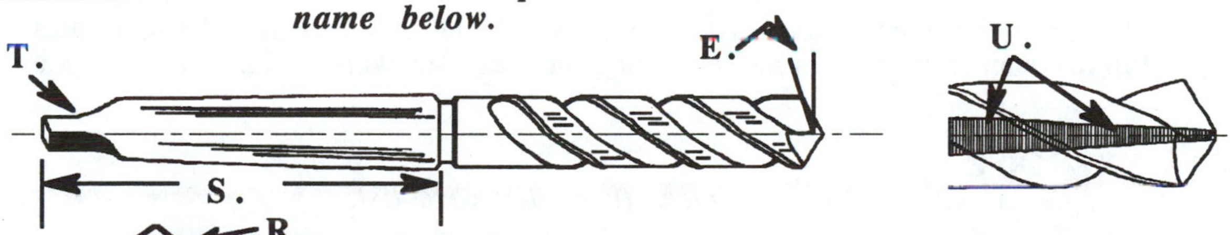

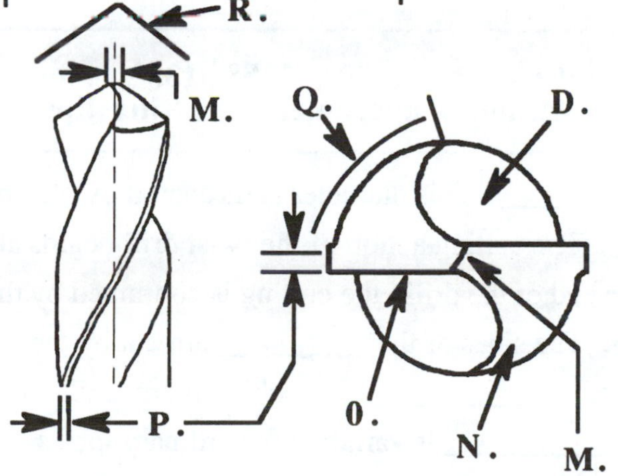

_____ 1. lip relief angle

_____ 2. cutting edge

_____ 3. chisel edge

_____ 4. margin

_____ 5. taper shank

_____ 6. heel

_____ 7. web

_____ 8. tang

_____ 9. point angle

_____ 10. cutting lip

_____ 11. body clearance

Directions: *Read the following paragraphs.*

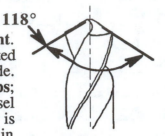

118°

regular
point angle

The study of twist drill parts continues with the parts located at the **point**. The first part of the drill to contact the workpiece is the **chisel edge** located at the tip of the twist drill; it is not a single point, but an edge like a blade. Running out from the chisel edge and angling down are the **cutting lips**; these lips (also called **cutting edges**) continue the cuts begun by the chisel edge. Looked at from the side, the downward angle of the cutting lips is called the **point angle**. The *regular point angle* is about 118° for use in ordinary drilling jobs; the *flat point angle* is about 135° for harder materials; and the *long point angle* is about 90° for drilling abrasive materials.

Behind the cutting edge, the drill material slopes down to the **heel**. This slope is provided to give room, or clearance, for the material being cut to go as it slides away into the flutes; the angle of the slope is about 8° to 12°, and is called the **clearance angle** or the **lip relief angle**.

The machinist wants to make a hole of a certain diameter. The **diameter of the drill** is what the size of the hole will be; this diameter is measured near the point with a micrometer which is placed around the margins of the drill. The **margins** are sharp edges that run the length of each flute down the body; the margins continue the work of cutting by removing material from the sides of the hole. To give clearance for this cutting action, the **body clearance**, located behind each margin, is made with a smaller diameter than the drill diameter.

If the margins are worn away, the drill will bind, because there is no clearance, even if the cutting edges have been resharpened. Regrind the margins wherever they are worn.

Do not trust the size stamped on the shank of a twist drill; mike the drill to learn its actual size.

Down through the center of the twist drill is a tapered area called the **web**. This is the solid metal core of the drill that supports the flutes and the cutting surfaces. The web is thinner near the point.

10-20:   **EXERCISE**
Directions: *Fill in the blanks with words from the word list.*

Word List:	chisel edge, margins, cutting, angle, 90°, 118°, 135°, hole, flutes, point, diameter, clearance, micrometer, cutting lips

The size of a drill is measured by the drill _____. This diameter is measured by placing a _____ around the _____. The cutting action of the twist drill begins at the _____ which is located at the end of the drill; the cutting is continued by the _____, also called _____ edges. The sides of the _____ are cut by the flute margins. All cutting edges must have some kind of _____ behind them to allow the material to flow into the _____. The point _____ is variable: for ordinary jobs a regular _____ angle is about _____; a flatter point is used for harder materials and has a point angle of about _____.

10-21:   READING
Directions: Read the following paragraphs.

DRILL SIZES

It is important to choose the right drill size. The shop drawing will give you a "call out" of the diameters of the holes you need to drill; these dimensions will be given in decimal fractions or in bar fractions. You will want to choose a drill the correct size for making the hole.

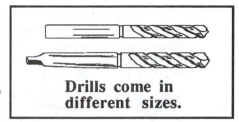

Drills come in different sizes.

Table H-1 (on pages 360 and 361 of the textbook) shows a list of drill sizes in **decimals of an inch** (to four decimal places) which range from .0135 in. up to .9844 in. To the right of these decimals are three columns which give you the names by which various drill sizes are known. There is a column which gives drill sizes in **bar fractions** ranging from **1/64** up to **63/64**. The next column gives drill sizes shown by **numbers,** with **80** (equal to .0135 in.) the smallest and **1** (equal to .2280 in.) the largest. After a size 1 drill, the column continues with **capital letters** as names for the rest of the drill sizes in that column (**A** = .2340 up to **Z** = .4130). The last column gives sizes in **millimeters** (the smallest is **.4 mm** = .0157 in.; the largest given is **25 mm** = .9843 in.). Here is a part of *Table H-1*:

Decimals of an inch	Inch	Number Letter	Millimeter
.2280		1	
.2283			5.8
.2323			5.9
.2340		A	
.2344	13/64		
.2362			6.0

Imagine that you pick up two twist drills and want to know what sizes they are. You look on the shank of the first, but the size has been worn off. You measure the drill diameter (at the margins near the point) with a micrometer and get a reading of .2344. You look at the table and see that this drill is a size 13/64. Next you pick up a new-looking drill that is stamped "A." You mike it and get a reading of .2341; you look in the table and find .2340 for A; at .2341 you are satisfied that this drill is correctly marked.

10-22:   EXERCISE:
Directions: Look at the drill-size table on pages 360 and 361 of your textbook. Use the table and write down drill sizes for these decimal inches:

Decimals of an inch	Drill size	Decimals of an inch	Drill size
.0313	1/32	.0595	_____
.0394	1 mm	.2420	_____
.0400	60	.2638	_____
.0550	_____	.3906	_____
.1250	_____	.4040	_____

Directions: *Look at these pictures. Listen to the names on the tape. Write the letters of what you hear next to the numbers.*

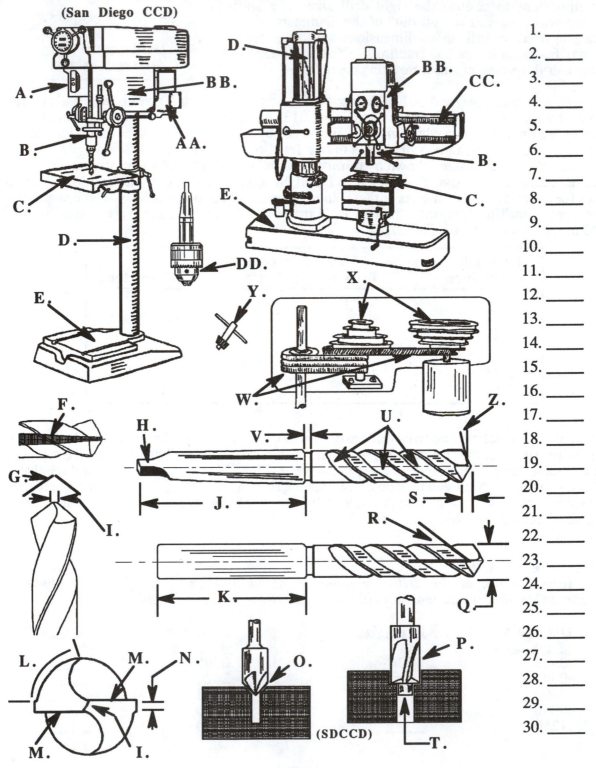

(San Diego CCD)

(SDCCD)

1. _____
2. _____
3. _____
4. _____
5. _____
6. _____
7. _____
8. _____
9. _____
10. _____
11. _____
12. _____
13. _____
14. _____
15. _____
16. _____
17. _____
18. _____
19. _____
20. _____
21. _____
22. _____
23. _____
24. _____
25. _____
26. _____
27. _____
28. _____
29. _____
30. _____

Unit 11: *TURNING MACHINES*

ASSIGNMENT: **Read, study and complete pages 119 to 142 of this book. Then read from pages 389 to 515 in the textbook, *Machine Tool Practices, SECTION I*, as directed by your teacher.**

OBJECTIVES for this unit:
You should be able to:
1. Identify and say correctly the important parts of the engine lathe.
2. Identify the names of a variety of lathe cutting tools and be able to name the parts of a toolbit.
3. Identify and pronounce the names of some important work-holding devices for the lathe.

11-01:   VOCABULARY LIST   A
Directions: Study the vocabulary.   Write the missing words in the blank spaces.

1. stationary

not moving, fixed in one place. *Example:*   In drilling, the workpiece is _____ ; in turning, the workpiece is rotated by the spindle.

2. toolbit

the cutting tool used in a lathe; it cuts material with a leading point at the end of the tool; it is also called a *single-point tool*. *Example:* The machinist usually grinds the point on each _____ he/she uses.

toolbit for use in a lathe

3. turret

a tool-holding device which is able to rotate in a circular path to present a variety of toolbits at the point of cutting. *Example:* Some lathes have a _____ with four or more tools in a circular pattern; each toolbit can be rotated into position for cutting.

4. to present

to put forward for use. *Example:* A turret is able _____ four or more toolbits, one at a time, for cutting a part.

5. automated

self-operated by means of computer programming. *Example:*   Lance's work produced more parts with the new, _____ equipment.

6. to slide

to move in a direction while touching a smooth surface. *Example:* Oil and grease help parts _____ over each other easily.

7. cast iron

a very hard form of iron which has been melted and formed in a desired shape. *Example:*   The bed of the lathe is made from _____.

8. to transmit

to send from one place to another. *Example:* When the lathe spindle turns, it is able _____ that rotation to the workpiece.

9. to aim

to point in a particular direction. *Example:* Victoriano learned how _____ the toolbit at the correct place on the workpiece.

10. possibility

an activity which can be done in the near future. *Example:* The operator saw the _____ of using a different tool to make the required cut.

119

Directions: Read the following paragraphs.

A LATHE ROTATES THE WORK

The drill press, which you studied in the last unit, is used to drill holes; it drives a rotating, cylindrical cutting tool, the twist drill, into a stationary workpiece. Now you will study another major machine, the **lathe**.

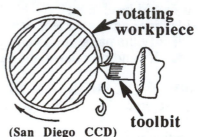

A frequently used type is the **engine lathe**. It is used to rotate a workpiece and then to move a single-point tool into the work, in an operation called **turning**. The cutting tool is called a **toolbit**; it has cutting edges running back from the point; they remove chips of metal as the tool cuts into the circumference of the spinning work. The engine lathe is the lathe most frequently found in school machine shops.

(San Diego CCD)

There are several other kinds and sizes of lathes which are used for a variety of machining tasks. Pages 385 to 393 in the textbook describe some of these special-purpose lathes. In addition to the engine lathe, you should be aware of a class of lathes used for mass production, the **turret lathes**; some of these are shown in Figures I-6, I-7, I-8, and I-11 in your text.

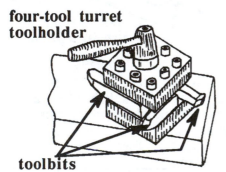

A **turret**, in machining, is a block or circle of metal which holds several different cutting tools; the turret rotates to present its different tools to the work. In this way a lathe, using a turret, can do different operations on the workpiece without stopping the machine to change cutting tools.

Sometimes lathes are **fully automated** and use CNC programming; a turret lathe of this variety can produce a completed product in a short time.

You measure the **size** of a lathe by the diameter of a cylindrical workpiece which the lathe can rotate around its axis above the ways; this diameter is called the **swing (D)**. A **radius** is a half swing **(R)**. The **maximum distance** between the two centers is called the **capacity** of the lathe **(C)**. The **length of the bed** is also shown in the drawing at the right **(B)**. An example of a commonly used shop lathe might have a swing of 13 inches and a capacity of 36 inches.

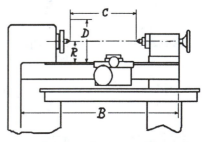

size and capacity

11-03: EXERCISE
Directions: For each statement, circle T for "true" or F for "false."

T　F　1. Turning is the operation of a lathe when it feeds a toolbit into a rotating workpiece.

T　F　2. Some CNC turret lathes can produce a completed part by using the tools in the turret.

T　F　3. A turret lathe is valued because it reduces the time needed for tool changes.

T　F　4. The engine lathe is the best machine for mass producing parts.

T　F　5. Work with a 7-inch radius can be turned on a lathe with a 13 inch swing.

The Parts of the Engine Lathe

Below are pictures of the engine lathe. The main parts are the **headstock**, the **carriage**, the **tailstock**, the **bed**, the **quick change gear box**, and the **base**. Round workpieces can be mounted between the headstock and the tailstock; the carriage allows the machinist to position the toolbit which is fed into the work. The feed can be changed by controls in the gear box.

 Directions: Study these pictures as you listen to the pronunciation of the parts of the engine lathe. Pronounce each part's name.

MAIN PARTS of THE LATHE

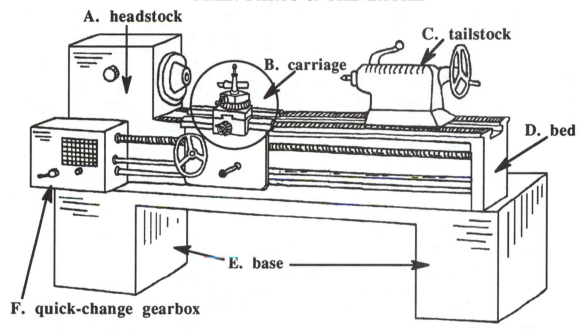

A. headstock

B. carriage

C. tailstock

D. bed

E. base

F. quick-change gearbox

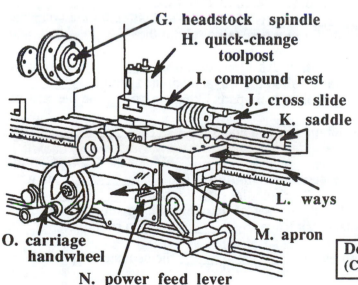

G. headstock spindle
H. quick-change toolpost
I. compound rest
J. cross slide
K. saddle
L. ways
M. apron
O. carriage handwheel
N. power feed lever

CARRIAGE FEATURES

The carriage slides along the **ways** in the right and left directions; this movement is controlled by the **carriage handwheel** or the **power feed lever**. The **cross slide** moves on a dovetail in and out for cross feed. The **compound rest** can be rotated to any angle to position the toolbit correctly.

Details of the Carriage
(Clausing Corporation)

121

TOOLPOSTS, OLD AND NEW

Some machine shop classrooms have a few older engine lathes; they use a **ring-and-rocker toolpost** to hold tools, as shown below and in the picture on the top of the previous page. The **toolbit** is secured in a **toolholder** which fits through a slot in the toolpost and rests on a **rocker** which can be tilted up or down several degrees in a concave **ring** at the bottom.

In today's machine shops you will probably see a more modern tool-holding device: the **quick-change toolpost**. This toolpost has a dovetail onto which a **toolholder** will quickly slide; the toolholder is locked in place by turning the **clamping handle**. Pages 406 and 407 in the textbook show the range of cutting tools that can be used with this toolholding device.

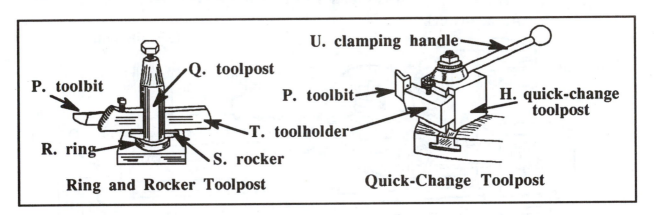

Ring and Rocker Toolpost

Quick-Change Toolpost

11-05:   CONVERSATION

Directions: *During the last week, Al Lopez, the instructor, has been teaching the students about the engine lathe. Lucy, Le, and Omar have come to class for a review. Listen to the conversation and fill in the missing words. Then rewind and listen again. Practice your pronunciation by listening and repeating.*

Al: Students, let's review what you've been learning about the _____ lathe. I'll give you one of the _____ parts of the lathe and you tell me two or three things you've learned about it. Okay, Lucy, tell us about the _____.

Lucy: The _____ is one of the main parts: It's a long cast iron _____ that runs the length of the lathe. It has a track, called the **ways**, machined onto its _____ surface. The bed _____ other main parts like the carriage and the tailstock.

Al: Good job, Lucy. Now, Le, let me ask you about the _____.

Le: The **headstock** is located on the _____ side of the lathe. Coming out from the headstock is the **headstock spindle**. The _____ is where you attach the _____. The spindle is what transmits _____ to the workpiece. The spindle sits in a **set of bearings** which support the spindle and _____ it in place, but leave it _____ to rotate.

Al: Good, Le. Can you also tell us what _____ the headstock spindle?

Le: Yes, inside the headstock are _____ or a belt-and-pulley system which drive the headstock spindle.

Al: Very good. And, Omar, now let's go on and _____ what you know about the _____.

Omar: The **tailstock** is one of those parts that Lucy mentioned which can move along the _____. The tailstock is located on the _____ side and can be moved toward the headstock. It also has a _____; the spindle can hold a pointed cylinder called a **dead center**, because it does not spin the work; its pointed _____ fits into one end of the workpiece, supporting it, and allowing it to _____, but not supplying any of the rotation.

Al: Yes, Omar; you've answered well. Now, Lucy, tell us what other main part can move along the _____.

Lucy: Teacher, the **carriage** is another main part that moves along the ways. The purpose of the _____ is to allow many angles for feeding the _____ into the work. The **toolbit** is held in a **toolholder** which _____ into a **tool post**. The _____ is attached to the **compound rest** which is like a turret that can turn in any _____ to aim the toolbit at the correct _____ toward the work.

Al: That's very good information, Lucy. Thank you! Now let's get Le to tell us more about the _____.

Le: Well, the compound rest is attached to the top of the **cross slide** which can _____ in and out across the ways. The cross slide itself _____ on top of the **saddle** which is the part of the carriage which slides to the _____ or to the left along the ways. That gives the _____ many possibilities for positioning the toolbit correctly. The last _____ of the carriage is the **apron**; it is a shield in front of some of the _____ parts. It contains gears that transmit _____ to the carriage. The carriage_____ is located on the apron.

Al: And what does that handwheel do?

Le: With the **carriage handwheel** the operator controls the movement of the whole carriage along the ways. With it the operator controls the _____ of the toolbit into the work. There is also a **power feed lever** on the apron; it sends the toolbit into the work at a rate the machinist sets.

Al: Good answers, Le. Now, Omar, wrap this review up by telling us about the **quick-change gearbox** and the **base**.

Omar: The _____ is used to make the lathe _____ on the floor. It also contains the motor that powers the lathe. The _____ contains gears that transmit _____ to the carriage for its feed motions.

Al: Good job. You all did well. Now let's say those six main parts one more time:
1) _____, 2) _____, 3) _____,
4) _____, 5) _____, and 6) _____.

123

11-06: EXERCISE
Directions: *Read the questions. Circle the **best** answer.*

1. The maximum diameter of a cylindrical workpiece which the lathe can rotate around its axis is
    called the:
    a. half swing   b. swing   c. capacity   d. lathe length

2. Which statement best describes the cutting action of a lathe?
    a. A rotating tool is pressed down into a stationary workpiece.
    b. The teeth of a cutting blade each remove material on the forward stroke of the blade.
    c. A workpiece is raised on a table into the rotating teeth of a cutting tool.
    d. A single-point cutting tool is fed into a rotating workpiece.

3. The maximum distance between the two centers of a lathe is called the:
    a. half swing   b. swing   c. capacity   d. lathe length

4. This main part of a lathe is located on the left side and has a tapered spindle for holding
    workpieces while they are rotating. It's a:
    a. headstock   b. bed   c. tailstock   d. carriage

5. What is shown in the picture to the right?
    a. a cast iron bed on which the carriage rides.
    b. a headstock spindle which is held by bearings.
    c. a toolbit with its single point for cutting.
    d. a toolholder which can hold up to four toolbits.

6. This main part of a lathe is made of cast iron; it has tracks, called *ways*, cut into its top
    surface. It's a:
    a. headstock   b. bed   c. tailstock   d. carriage

7. In addition to the saddle, what other part can move along the ways?
    a. headstock   b. bed   c. tailstock   d. quick-change gearbox

8. What part actually cuts into the rotating workpiece?
    a. toolbit   b. tool post   c. toolholder   d. carriage handwheel

9. Where is the carriage handwheel located?
    a. on the saddle   c. over the compound rest
    b. on the apron   d. at the rear end of the tailstock

10. What main part transmits power to the carriage for its feed motions?
    a. the headstock spindle   c. the base
    b. the tailstock spindle   d. the quick-change gearbox

Extra Question: Write the names of the six main parts of the engine lathe from page 121.

      1. _____   4. _____

      2. _____   5. _____

      3. _____   6. _____

The Parts of the Engine Lathe

Directions: *Continue learning the names of the engine lathe parts. Look at the pictures. Then write the correct letter by each name below.*

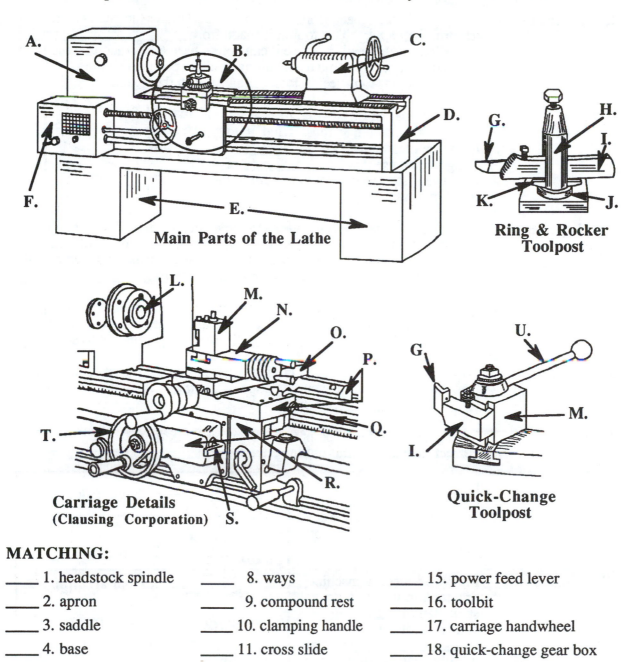

Main Parts of the Lathe

Ring & Rocker Toolpost

Carriage Details
(Clausing Corporation)

Quick-Change Toolpost

MATCHING:

____ 1. headstock spindle	____ 8. ways	____ 15. power feed lever
____ 2. apron	____ 9. compound rest	____ 16. toolbit
____ 3. saddle	____ 10. clamping handle	____ 17. carriage handwheel
____ 4. base	____ 11. cross slide	____ 18. quick-change gear box
____ 5. toolholder	____ 12. toolpost	____ 19. quick-change toolpost
____ 6. ring	____ 13. rocker	____ 20. carriage
____ 7. headstock	____ 14. tailstock	____ 21. bed

11-08: VOCABULARY LIST B

Directions: *Study the vocabulary. Write the missing words in the blank spaces.*

1. advice — an opinion or idea, given to another person, about what to do in a particular situation. *Example:* The teacher can give you some good _____ about what to do, if you have a problem with running the lathe.

2. high speed steel — a very hard metal made from iron, carbon, and either tungsten or molybdenum. *Example:* This toolbit is made from _____; it's called "high speed" because it can not be harmed as much by the high temperatures which come from cutting at high speeds. For more information on "tool steels," look at page 198 in the textbook.

3. blank — the small bar of metal used to make a toolbit; the machinist must grind it to make the tool that is needed; *Example:* I'm going to use this _____ of high speed steel to make a toolbit for my roughing cuts.

"blank" for making a toolbit

4. tool geometry — Knowledge of the correct angles and dimensions which should be ground on cutting tools in order to make efficient cuts and to avoid tool wear. *Example:* Andrea has shaped many toolbits on the grinder; she knows the correct angles and is very good with all the parts of _____.

5. task — a piece of work given to a worker by another person; a job which requires study and hard work. *Example:* Learning the names and purposes of the machine tools is just one _____ which you must do to succeed.

6. clearance — grinding an angle behind any cutting edge, so that the material being cut can flow more easily past the cutting edge. *Example:* Another word for _____ is *relief.*

7. keen — having a sharp edge or point that will cut well. *Example:* Because of correct rake angles, this cutting edge is very _____.

8. jagged — having sharp, uneven points. *Example:* When the point of toolbit broke, there was a _____ edge left on the toolbit.

9. to chatter — for a workpiece, a machine, or a tool, to vibrate or to shake rapidly. *Example:* When a tool begins _____, you can usually hear the vibration and see wavy marks on the workpiece.

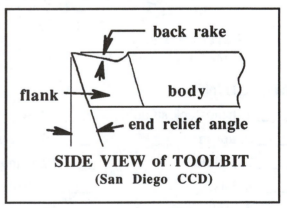

SIDE VIEW of TOOLBIT
(San Diego CCD)

THE SHAPE OF LATHE TOOLBITS

You will be able to grind your own lathe toolbits when you learn about the shape and the angles of a toolbit. Usually you will be doing this to a piece of high-speed tool steel called a "blank." Your task for now is to learn the names of the parts and angles of the toolbit. Then you will be able to ask questions and describe problems you may have with this "tool geometry." The three views given below show the important parts and angles of a roughing tool.

Shown in the top view are two cutting edges, the **end cutting edge** and the **side cutting edge**; the two edges meet to form the **point angle**; however the point angle does not have a sharp point, because that point would easily break with use; instead, the point is given a rounded shape, called the **nose radius**. In your textbook, Table 1-1 gives 1/32 inch for the nose radius; that means that all points on the rounded end are equidistant from a point set in 1/32 inch from the end. The nose and the two cutting edges receive the most pressure during cutting.

Both cutting edges are given **clearance** by grinding relief angles behind them: the end cutting edge is given **end relief**, as shown in the side view, and the side cutting edge is given **side relief**, as shown in the end view. By grinding these relief angles, the edge of the tool will be freed from binding (binding is when the tool cannot move because of too much material); the material can then flow freely past the cutting edges as the chips curl outward from the workpiece.

Rake angles are also ground on the toolbit to give the best cutting angles to the two edges: the **back rake**, as seen in the side view, helps with smooth chip flow and good finish; the **side rake**, as seen in the end view helps with chip flow and provides a keen edge.

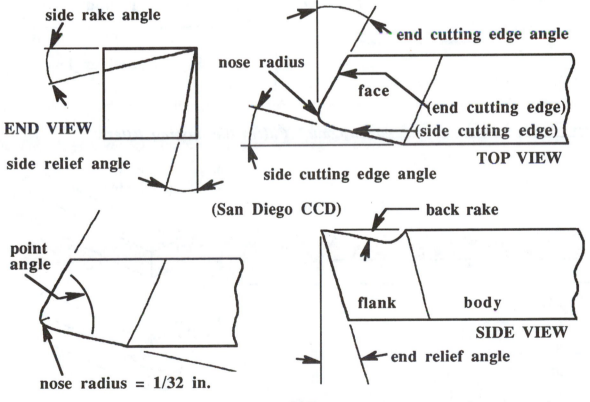

127

The Parts and Angles of a Toolbit

Directions: Study these pictures as you listen to the pronunciation of the nomenclature used in toolbit geometry. Pronounce each term. Rewind, listen again and complete the exercises which follow:

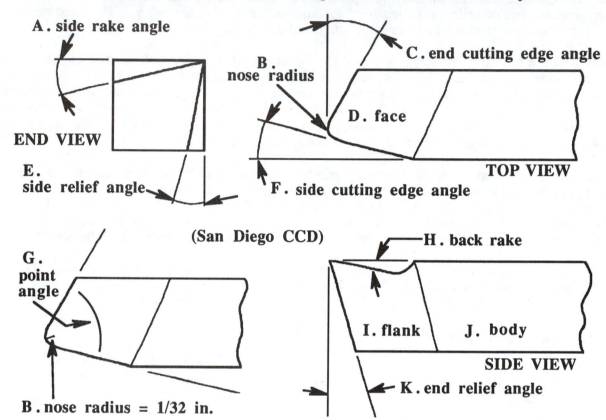

A. side rake angle

END VIEW

E.
side relief angle

B.
nose radius

C. end cutting edge angle

D. face

TOP VIEW

F. side cutting edge angle

(San Diego CCD)

G.
point
angle

B. nose radius = 1/32 in.

H. back rake

I. flank   J. body

SIDE VIEW

K. end relief angle

SPELLING: *Listen to the words again. Fill in the missing letters.*

A. s _ d _   r _ _ e a _ g _ e

B. _ o _ e r _ _ _ i _ s

C. e _ d _ u _ _ in _ ed _ e _ n _ le

D. _ a _ e

E. s _ _ _ e r _ l _ _ _ f _ _ g _ e

F. _ i _ e c _ t _ in _ _ d _ e _ n _ _ e

G. p _ i _ t _ n _ l _

H. _ ac _ _ a _ _

I. f _ a _ _

J. _ od _

K. _ n _ r _ li _ f _ n _ le

blank made of
high speed steel

finished
toolbit

11-11: "SHOP TALK"   STATING A PROBLEM & GETTING ADVICE

Directions:
You can use your knowledge of toolbit nomenclature to describe problems and to understand your teacher's advice. Listen to these dialogues, fill in the missing words, then rewind and practice the *pronunciation*.

1. Student's problem:

I've been _____ this aluminum pulley for a sailboat, but I'm not very happy with the finish; it's pretty _____.

   Teacher's advice:

Aluminum is a _____ metal, and your toolbit needs some back _____. Make sure you have about _____ of back rake.

2. Student's problem:

Look at this workpiece. See, the finish is very rough and there are deep _____ in the work.

   Teacher's advice:

As I look at your toolbit, I notice the point has broken off and left a jagged, uneven _____. Regrind it, and this time, put a nose _____ of _____ inch on the end.

3. Student's problem:

The toolbit is _____ as it cuts the material.

   Teacher's advice:

Sometimes chatter occurs when the nose radius is too large. You can try grinding a _____ radius.

4. Student's problem:

My friend is working at a shop that recently got an order to make ten thousand parts like this one. He's not sure what _____ to _____ for such a big job, and asked me to ask you for some _____.

   Teacher's advice:

Shops like where your friend works usually have smaller orders and use _____ _____ steel tools for the job. For an order of ten thousand he may want to use tools with _____ tips. They'd last longer.

5. Student's problem:

I'm getting one long, unbroken _____ that's tangled and very sharp.

   Teacher's advice:

You may want to _____ another toolbit with a smaller angle for the _____ rake; that will _____ the chip more and it will break more easily. You should also grind a chip breaker on the tool. The chip breaker will curl the chip back into the work and help _____ it off. Your best chip looks like a figure _____.

129

11-12: EXERCISE

Directions: *For each statement, circle T for "true" or F for "false."*

T   F   1. The length of a nose radius on a toolbit is always 1/32 in.

T   F   2. The back rake and the end rake on a toolbit are made to give a keen cutting angle.

T   F   3. High speed steel toolbits are more often used in mass production than are carbide.

T   F   4. The best chip produced by the lathe is long and unbroken.

T   F   5. "Tool geometry" is about the angles and lengths needed for an efficient cutting tool.

11-13:   TOOLBIT NOMENCLATURE

The Parts and Angles of a Toolbit

Directions: *Study these pictures of the parts of a toolbit; then find the name for each part from the list below. Match the letters with the numbers.*

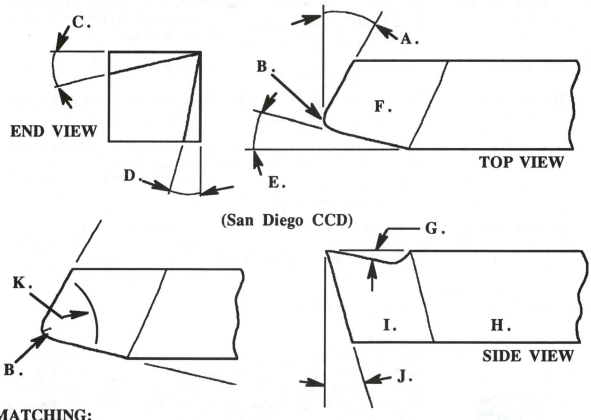

(San Diego CCD)

MATCHING:

____ 1. nose radius   ____ 5. side rake angle   ____ 9. end relief angle

____ 2. flank   ____ 6. end cutting edge angle   ____ 10. face

____ 3. side relief angle   ____ 7. body   ____ 11. point angle

____ 4. back rake   ____ 8. side cutting edge angle

11-14: VOCABULARY LIST C
Directions: *Study the vocabulary.* *Write the missing words in the blank spaces.*

1. shape — the external surface of something; the exterior form of something. *Example*: With a lathe, a machinist can cut the sides of a cylinder to make the _____ of a screwdriver.

shapes

2. to perform — to bring to completion; to do a task. *Example*: Among the tasks that the machinist has _____, is the job of cleaning and oiling the machines.

3. to part — to cut or break into two or more pieces. *Example*: Maribel used a special tool _____ the round stock into three pieces.

4. concave — curving inward like the inside surface of half a rubber ball. *Example*: The inside of this bowl is _____.

concave surface

5. convex — curving outward, like the outside surface of half a rubber ball. *Example*: The outside surface of an orange or an apple is _____.

convex surface

half a rubber ball

6. juncture — a point, line, or area where things join or connect. *Example*: The teacher told me that a fillet is the concave _____ of two surfaces.

7. shoulder — the part of a cylindrical object formed by a face and a side. *Example*: That _____ has a sharp edge; you should chamfer it.

shoulder

8. groove — a long, narrow hollow cut into the surface of a workpiece. *Example*: Here are two examples of a _____; one of them is rounded, and one of them is square.

rounded groove

square groove

grooves

9. to reduce — to lessen in any way, as in size, length, or weight. *Example*: Aki was able _____ the diameter by turning the workpiece.

SAFETY TIPS:

1. *Carry sharp tools with a shop rag to avoid cutting your hands.*
2. *Wear heavy shoes to protect your feet if you drop a heavy shop tool.*
3. *When someone wants to hand you a heavy tool, watch out for your hands and feet.*

Directions: Read the following paragraphs.

WORKPIECE SHAPES AND TOOL SHAPES

The purpose of turning round stock is to give certain shapes to the metal. In the drawing below are certain more common shapes which the machinist can produce. Study and be able to use these words for shapes. Learn also what tools are used to make the shapes or perform other tasks.

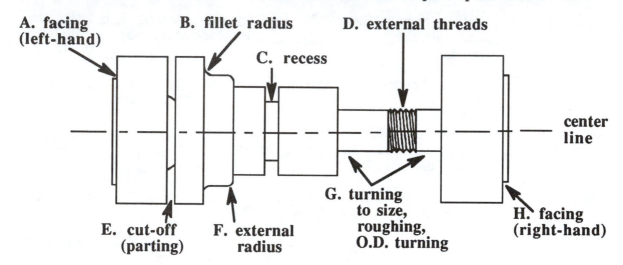

A. facing (left-hand)
B. fillet radius
C. recess
D. external threads
E. cut-off (parting)
F. external radius
G. turning to size, roughing, O.D. turning
H. facing (right-hand)
center line

1. Facing: Accuracy in the final parts requires that the two ends of the round stock used in the lathe have very flat ends. Making each end flat by the removal of small amounts of material from the two ends is called **facing**. The tool used for facing is called a **facing tool**; there are two kinds: the **left-hand** facing tool (which cuts toward the tailstock) and the **right-hand** facing tool. (which cuts toward the headstock).

l. h. facing tool

r. h. facing tool

parting tool

2. Cut-off: Round stock comes in longer lengths, and often a shorter piece must be cut off before or after the part has been worked on. This cutting of the stock or part into two pieces is called **cut-off** or **parting**. The machinist will either saw the piece or will use a **parting tool**; a parting tool is narrow (not wide) and is used to cut the work in two.

3. Fillet radius: A **fillet** is a concave juncture of two surfaces. The juncture is curved and so will have an inside corner radius. A **fillet radius tool** can be made to cut whatever size fillet is needed.

fillet radius tool

4. External radius: A **shoulder** is the right-angled edge of a part that is left after material has been removed by a tool cutting at right angles to the center line. Such a shoulder could be chamfered, or it could be given an **external radius**, an even rounding of the shoulder. An **external radius tool** can be used for this.

external radius tool

132

11-15: READING (continued)

recessing tool

5. Recess: A recess is a groove which has been cut into a piece of round work at right angles to the center line of the work. A **recessing tool** is used for this.

6. External Threads: Often a cylindrical workpiece needs **external threads**, because another part with mating internal threads will be joined to it. Smaller external threads can be cut with dyes, but they are commonly cut on the lathe, using a 60° **threading tool** (the point angle of the tool is 60°).

threading tool

7. Turning to size: The **roughing tool** you studied in the previous pages can be used to remove any areas on a cylindrical workpiece that need to be reduced in diameter. You can cut any outside diameter that is needed. The roughing tool can be **right hand** (cutting toward the headstock) or **left hand** (cutting toward the tailstock.)

l.h. roughing tool

r.h. roughing tool

11-16: EXERCISE

Directions: Look at the numbered cuts on this workpiece. Then look at the names in the "Word List." Write the correct word after each number:

> **Word List:** recess, left-hand roughing, external radius, cut-off, external threads, right-hand facing, fillet radius, right-hand roughing, left-hand facing.

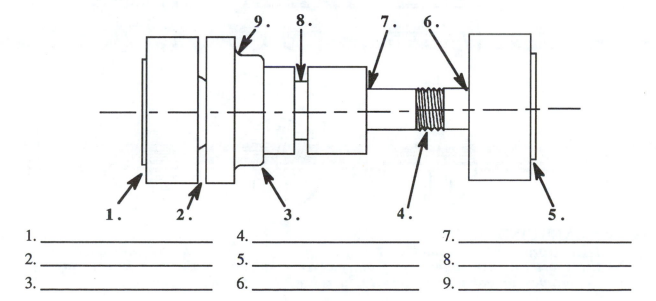

1. _____   4. _____   7. _____

2. _____   5. _____   8. _____

3. _____   6. _____   9. _____

Directions: *Listen to the tape as the student talks about some job situations. The teacher will offer three suggestions for each job. Circle the letter of the best advice.*

Job Situation   **Teacher's advice**

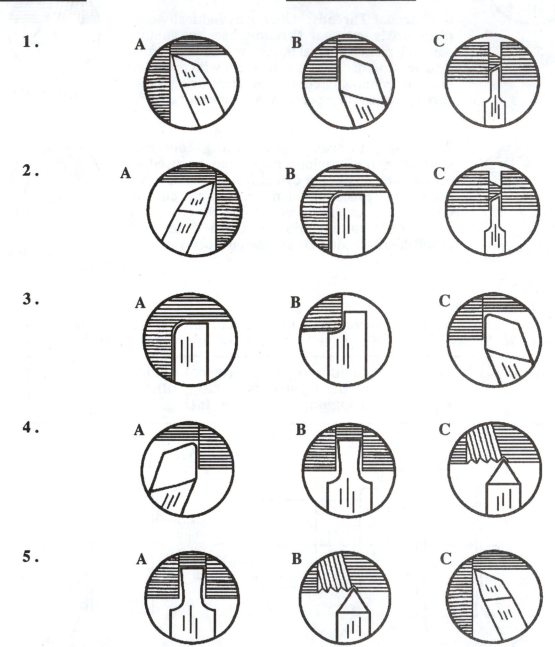

11-18: EXERCISE

Directions: *As a second exercise, see how fast you can say the names of the tools on this page: Go from picture to picture with your finger, saying the names of each tool. Try to increase your speed.*

11-19: VOCABULARY LIST D
Directions: Study the vocabulary. Write the missing words in the blank spaces.

1. to drive — to transmit motion to a machine part or to a workpiece. *Example:* The motion of the spindle is able _____ the chuck which holds the workpiece.

2. to insert — to put into. *Example:* Don't forget _____ your time card in the time clock slot; the clock prints the time of your arrival.

3. Morse taper — a taper is a gradual decrease in width, thicknes, or cross-section diameter moving from one end to the other; the Morse taper decreases 5/8 inch in diameter for each foot of length. *Example:* The spindle in this lathe has a _____ _____ into which mating shanks can be inserted.

taper

(SDCCD)

notch →

4. notch — a concave cut in the edge or surface of an object. *Example:* A lathe drive plate has a _____ cut into one edge.

drive plates

5. stud — a cylindrical piece coming up out of a surface; it is used as a support, stop, or pivot. *Example:* Instead of a notch this drive plate has a _____ projecting from its surface.

stud

6. variation — change in form from what is usually seen. *Example:* A _____ of a ham sandwich has a layer of ham and a fried egg, instead of lettuce.

7. concentric — having a center in common. *Example:* Here are some _____ circles.
A related word is *concentricity:* it is an important quality in some workpieces; many machined cylindrical surfaces must be concentric; they must have concentricity.

concentric circles

8. eccentric — having a center different from the common center. *Example:* When a circular feature does not have a center in common with other circular features, it is said to be _____. In the picture, the two small circles have a common center, C, but the larger circle has its own center, P; it is eccentric with respect to C.

9. independent — operating separately, without needing something else to happen. *Example:* If each jaw of a chuck moves separately and not together, the chuck is said to be _____.

10. irregular — not regular; not having the same shape all the way around. *Example:* This part has an _____ shape, so it cannot be turned between centers.

135

11-20: READING
Directions: Read the following paragraphs.

HOW WORK IS HELD IN THE LATHE

Holding the work on a lathe requires two things: First, the work must be secure, because it will rotate at high RPMs and could be thrown across the room if it is not held tightly. Secondly, the work-holding device must be free to drive the work in a rotational path, so various cutting tools can be fed into it.

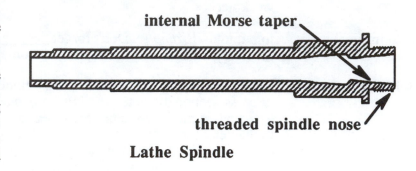

internal Morse taper

threaded spindle nose

Lathe Spindle

The **spindle** shown in this picture is located in the headstock; older models have a **threaded spindle nose** to which work-holding devices can be screwed directly. The **internal Morse taper** is more often used to attach work-holding devices.

Among the more important work-holding devices, which are described and illustrated in your textbook, are the following: lathe centers and drive plates, lathe chucks, face plates, collets, tapered mandrels, and steady rests. Study the examples given here and in the textbook.

1. Lathe centers and drive plates for turning between centers: "To turn between centers" means to mount a piece of round stock between a **spindle center** mounted in the spindle taper of the headstock and a **rolling center** or **ball bearing center** in the tailstock; then a single point tool is used to remove material from the rotating work. The two ends of the workpiece are faced and center-drilled; the two centers are fit into these two holes; the centers support the workpiece, but allow it to rotate freely.

Often a **drive plate** is mounted on the headstock in order to rotate the work; the spindle center has a tapered shank which is inserted through the center of this plate into the spindle and turns with it. The drive plate has a notch into which the **tail** of a **lathe dog** is placed. The lathe dog has an opening into which the work fits; the work is gripped by the lathe dog **set screw**. The spindle drives the plate with its lathe dog attached; the dog drives the work.

notch

set screw

hole for center

drive plate
(SDCCD)

bent-tail lathe dog
(SDCCD)

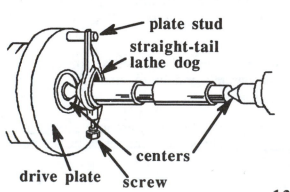

plate stud

straight-tail lathe dog

centers

drive plate   screw

In the picture to the right, a **straight-tail lathe dog** secures a cylindrical workpiece and is driven by a stud on the drive plate. This is a variation of the **bent-tail lathe dog** which fits into the drive plate notch. The work rides between two centers.

11-20: READING

For work made of soft metals, like aluminum, some-times a **drive center** can be inserted in the head-stock spindle. This center has grooves cut around the point's periphery which are forced into the end of the soft metal workpiece to hold and drive it. In that arrangement a drive plate is not needed.

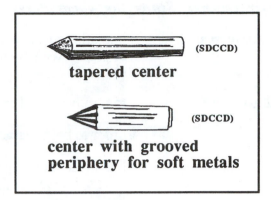

tapered center (SDCCD)

center with grooved periphery for soft metals (SDCCD)

It is important that the centers be **aligned** properly when turning between centers. The centers must lie along the same axis. Before beginning the final job, a machinist can use a test piece to make sure of the alignment of the centers. If the centers are "true" then all diameters cut by the lathe will be **concentric**. If the centers are not aligned properly, some of the cuts will have **eccentric** shapes, which means the work really has two centers which have two different locations.

11-21: EXERCISE
Directions: Read the questions. Circle "T" for True and "F" for False.

T F 1. The main spindle, as shown in this unit, is located in the tailstock.

T F 2. Some spindle noses have external threads to which work-holding devices can be screwed directly.

T F 3. Turning between centers requires a spindle center in the headstock and a rolling or ball bearing center in the tailstock.

T F 4. Drive plates are mounted on the spindle in the tailstock.

T F 5. One main way of holding work-holding devices is by the use of mating Morse tapers.

T F 6. A bent-tail lathe dog is rotated by a drive plate with a projecting stud.

T F 7. A straight-tail lathe dog uses a screw to secure the work which passes through it.

11-22: READING
Directions: Read the following paragraphs.

2. Independent lathe chucks: You are familiar already with how a drill chuck has jaws which can be tightened around a twist drill to hold it securely. The lathe often uses chucks to hold the workpiece at the end of the spindle nose. One kind is the **four-jaw independent chuck**; it is called "independent" be-cause each of the four jaws can be moved one at a time; this allows the chuck to hold irregular-shaped workpieces, as shown in Fig. I-113 in the textbook.

The four jaws of the independent chuck are closed one at a time, in order for them to hold an irregular workpiece. A **chuck key** is used to tighten the individual jaws.

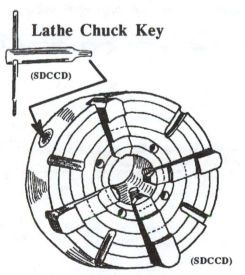

Lathe Chuck Key

(SDCCD)

(SDCCD)

Four-jaw Independent Chuck

137

11-23: VOCABULARY LIST E
Directions: Study the vocabulary. Write the missing words in the blank spaces.

1. universal   operating together, not independently. *Example:* When I turn the chuck key, all three jaws close around the work. This is a _____ chuck.

2. to reverse   to turn backward or in the opposite direction. *Example:* Sometimes it is necessary _____ the direction of the chuck jaws, in order to hold a larger workpiece.

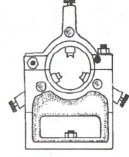

The spindle is hollow inside.

3. hollow   having an empty space inside. *Example:* This spindle is _____ inside; the _____ space has an internal taper called a Morse taper.

4. to project   to stick out from a surface. *Example:* The studs are made _____ from the drive plates.

5. steady   not moving; holding in the correct position. *Example:* Longer workpieces may need to be held _____ by one or more steady rests.

6. to sag   to sink, bend, or curve from weight or pressure. *Example:* A long, thin workpiece is going _____ unless it is supported by a steady rest.

> **A steady rest receives the work through the opening. The position of the work is adjusted by the three screws.**

steady rest
(SDCCD)

11-24: READING
Directions: Read the following paragraphs.

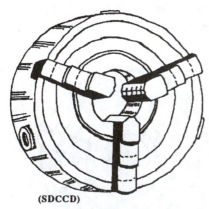

(SDCCD)
Universal Chuck

3. Universal lathe chucks: A second type of lathe chuck is the **universal chuck** which usually has three jaws; these three jaws are not independent and close around the workpiece at the same time, with all three jaws moving in or out an equal distance. These chucks are quick and easy for use with round stock. A separate set of jaws is available for holding larger diameter pieces.

The lathe chuck usually holds work which is relatively short. If the work is longer, it may be mounted between a chuck and the tailstock dead center which will support the other end. For long pieces with narrow diameters, a **steady rest** may be used to support the work and keep it from sagging.

11-24: READING (continued)

In the picture below a long piece of round work is held in a lathe chuck and is supported along its length by a steady rest to prevent sagging and to keep the work aligned. Adjustments in centering are made with the three screws.

a steady rest in position; it helps support the work and keep it aligned

Some chucks have jaws that can be reversed or which have two sets of jaws; this allows the machinist to do **outside chucking**, in which the jaws grip the outside of the workpiece, or **inside chucking** in which the jaws fit inside a hollow, cylindrical workpiece.

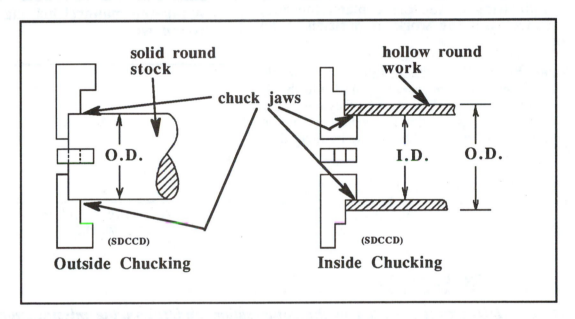

11-25: EXERCISE
Directions: Read the questions. Circle "T" for True and "F" for False.

T F 1. All four jaws of the independent chuck tighten around the work at the same time.

T F 2. The three-jaw universal chuck is often used to hold irregular shaped workpieces.

T F 3. A chuck key is used to tighten the individual jaws of an independent chuck.

T F 4. Hollow round work can be held by inside chucking.

T F 5. All work held in lathe chucks requires the use of a steady rest.

T F 6. A center with a grooved periphery can be used to both hold and drive soft work.

T F 7. In order to achieve a concentric workpiece, the alignment of centers is important.

T F 8. The only lathe dog that can be used with a drive plate is a bent-tail dog.

11-26: READING
Directions: Read the following paragraphs.

4. Tapered mandrels: Another work-holding device is the **tapered mandrel**. This lathe accessory comes in several sizes; each is a cylinder with a taper of .006 in. per foot. The mandrel is driven by a drive plate and lathe dog set up. It is used to hold round work that has a hole drilled in the center; the mandrel fits snugly through the hole in the work and helps position it along the ways where lathe tools can cut it.

There is more information about tapered mandrels on pages 449 to 451 of the textbook. Look at **Figure I-210** in the textbook; then, in the box at the right, **make a drawing of the drive plate, the dog, the mandrel and the work in position to be turned.**

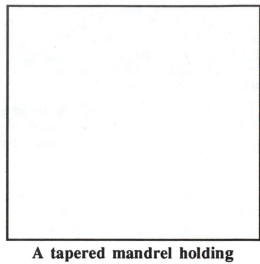

A tapered mandrel holding round work.

5. Collet chuck: Collet chucks are another work-holding device. They are especially useful for holding round work with small diameters; they come in a variety of sizes, so get one that fits well; it should not vary in size from the work it is to hold by more than +.002 to -.003 inch.

There is more information about collets on pages 425 to 416 of the textbook. **At the right, draw a picture of the spring collet shown in Figure I-128.**

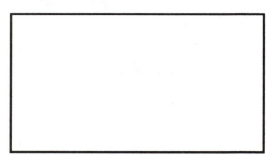

A spring collet for round work.

11-27: CONVERSATION

Directions: Listen to the conversation and fill in the missing words. Then listen again as you practice your pronunciation.

Al: Let's do a little reviewing. How many different work-holding _____ can you think of that are used with an engine _____?

Lucy: First off, there's drive _____ and lathe _____ that can be used to hold and drive work _____ centers.

Omar: I remember some different kinds of chucks: There's a four-jaw _____ chuck for irregular work, and there's also a three-jaw _____ chuck for turning _____ work between centers. Oh, and we recently learned about _____ chucks which are good for holding work with small diameters.

Lucy: We also learned about _____ mandrels: Round work that has a centered hole can be forced on a _____ with an arbor press. Then the mandrel is driven by a plate and a lathe dog during the turning.

140

Directions: Test yourself on the names of the parts that are used with lathes. Listen to the names on the tape. Write the letters of what you hear next to the numbers.

Part A: Lathe Main Parts, Carriage Details, and Toolposts:

1. ____
2. ____
3. ____
4. ____
5. ____
6. ____
7. ____
8. ____
9. ____
10. ____
11. ____
12. ____
13. ____
14. ____
15. ____
16. ____
17. ____
18. ____
19. ____
20. ____
21. ____

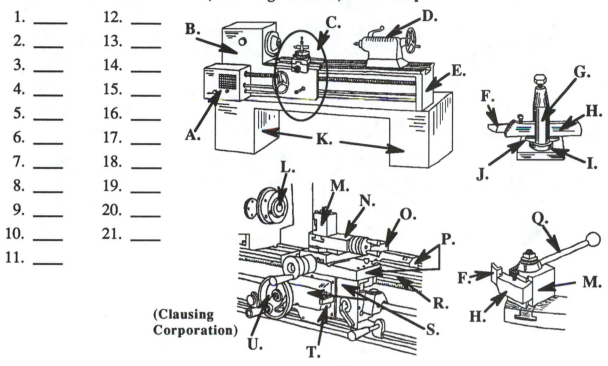

(Clausing Corporation)

Part B: Toolbit Features and Angles:

22. ____
23. ____
24. ____
25. ____
26. ____
27. ____
28. ____
29. ____
30. ____
31. ____
32. ____

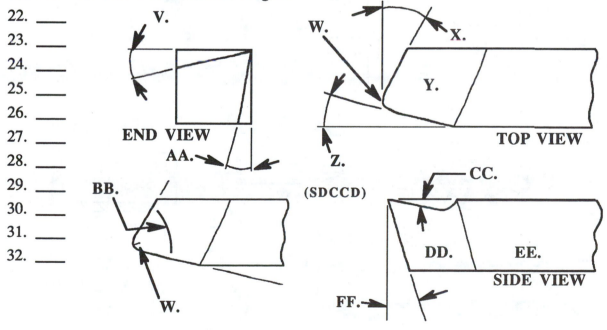

END VIEW

TOP VIEW

(SDCCD)

SIDE VIEW

141

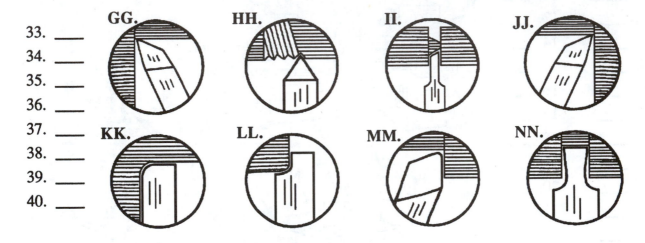

Directions: Test yourself on the names of the parts that are used with lathes. Listen to the names on the tape. Write the letters of what you hear next to the numbers.

Part C: Lathe Tools:

33. ___
34. ___
35. ___
36. ___
37. ___
38. ___
39. ___
40. ___

GG.   HH.   II.   JJ.

KK.   LL.   MM.   NN.

Part D: Lathe Part-Holding Devices:

41. ___
42. ___
43. ___
44. ___
45. ___
46. ___
47. ___
48. ___
49. ___
50. ___
51. ___
52. ___
53. ___
54. ___

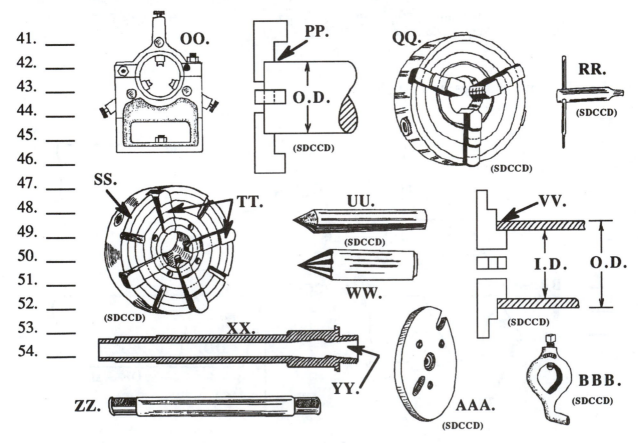

OO.   PP.   QQ.   RR. (SDCCD)

O.D. (SDCCD)   (SDCCD)

SS.   TT.   UU. (SDCCD)   VV.

WW.   I.D.   O.D. (SDCCD)

XX.

YY.   AAA. (SDCCD)   BBB. (SDCCD)

ZZ.

Unit 12: *VERTICAL MILLING MACHINES*

ASSIGNMENT: **Study and complete pages 143 to 162 of this *Guide*.**
Then read pages 516 to 548 in *Machine Tool Practices*,
SECTION J, Units 1 to 5, "Vertical Milling Machines."

OBJECTIVES for this unit:
The student should be able to:
1. Identify the important parts of the vertical milling machine and describe the functions of each.
2. Identify some controls on a vertical mill and describe the functions of each.
3. Identify common cutters for the vertical mill and tell for what operations each is used.
4. Identify common safety hazards for the mill and state safety rules for each.

12-01: VOCABULARY LIST A
Directions: Study the vocabulary and fill in the blanks with the missing words.

1. to feed	to move a workpiece into a cutter or a cutter into a workpiece. *Example:* The job of the table is _____ the part up into the spinning cutter.	
2. pocket	An open space cut into some surface of a part. *Example:* A vertical mill is good for cutting a _____.	
3. slot	a narrow groove cut into a part. *Example:* Use this cutter to make that _____.	
4. to chamfer	to remove the sharp edge of a new part by cutting it at an angle, usually a 45° angle. *Example:* Use this cutter _____ that edge.	
5. adapter	a part that helps one part fit into a holder of a different size. *Example:* Minh, use that _____ to fit this cutter into the spindle.	

12-02: READING
Directions: Read the following paragraphs.

Two Types of Milling Machines

Milling machines (frequently called mills) are useful, efficient tools. The cutting tools for mills are called **cutters**. During milling operations the workpiece is secured to a movable **table**. A part of the mill called a **knee** raises the table; the work is fed up into the rotating cutters. There are a variety of cutters which can remove metal in various ways from the workpiece: some cutters can cut pockets, some can make slots, others can chamfer edges, others can drill holes, others can cut special shapes into the metal.

There are two main types of milling machines: **vertical mills** and **horizontal mills**.

What is the difference between the two kinds of machines? Look at the axes of the machines. The words **vertical** and **horizontal** refer to the axis of the spindle on each machine.

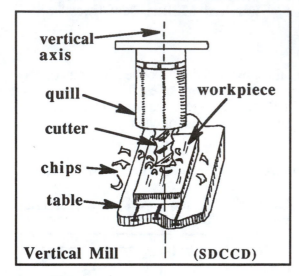

Vertical Mill (SDCCD)

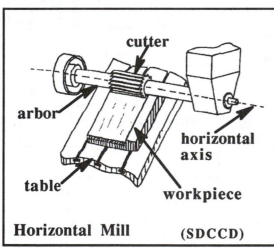

Horizontal Mill (SDCCD)

In a vertical mill the **axis** of the spindle runs vertically (up and down). The cutter on a vertical mill has a **shank** which fits into the spindle or into an **adapter** which will fit into the spindle. Because they have shanks, the cutters look like twist drills, but a vertical mill and a drill press are different: A drill press lowers a rotating drill into work which is secured to a stationary table; the vertical mill raises the work into a rotating cutter. Sometimes, however, the mill is used just like a drill press when it uses a tube called a **quill** to drill into the work in an up and down motion.

The study of the vertical mill will be the focus of this unit.

By way of contrast, on a horizontal mill the axis of the spindle runs from left to right in a horizontal plane. The cutters on a horizontal mill are round wheels or cylinders with holes in the middle and with cutting teeth on the periphery and sometimes the sides. To use a cutter a bar, called an **arbor**, is placed through the center hole of the cutter; one end of the arbor is inserted into a spindle which rotates the arbor and the cutter on it. Again, because it is a mill, a table moves upward to feed the work into the cutters.

The nomenclature of horizontal mills will be studied in the next unit.

12-03: EXERCISE
Directions: Read the questions. Circle the <u>best</u> answer.

1. For which operation can the vertical mill be used?
   - a. cutting pockets
   - b. making slots
   - c. chamfering edges
   - d. all of the above

2. The vertical mill is called *vertical*, because:
   - a. the axis of the mill's spindle is vertical.
   - b. the mill raises the work vertically into the cutter.
   - c. the mill can cut vertical slots.
   - d. none of these.

12-03: EXERCISE (continued)

3. What are the most frequently used metal-removing tools on a vertical mill called?
   a. twist drills
   b. toolbits
   c. cutters
   d. shanks

4. A mill which uses an arbor is called:
   a. a vertical mill
   b. a horizontal mill
   c. a shank mill
   d. an adapter mill

5. To remove a sharp edge of a part by cutting the edge at a 45° angle is to make a:
   a. pocket
   b. slot
   c. chamfer
   d. hole

6. A milling cutter was used to make this part. What is the best name of the opening at the arrow?
   a. pocket
   b. slot
   c. hole
   d. operation

7. An invisible line that passes through the center of a cylinder, a taper, or a ball is called:
   a. slot
   b. chamfer
   c. axis
   d. shank

8. Which machine uses a stationary table?
   a. drill press
   b. lathe
   c. horizontal mill
   d. vertical mill

9. When a vertical mill is used like a drill press, the operator makes the cut by moving:
   a. the arbor
   b. the quill
   c. the table
   d. the horizontal axis

12-04: VOCABULARY LIST B
Directions: *Study the vocabulary; write the missing words in the blanks.*

1. to support — to hold up the weight of something else; to keep from falling. *Example*: The base is made _____ the weight of the other parts.

2. to position — to put something in a place where you want it. *Example*: Carlos was very careful _____ the cutter in the center of the arbor.

3. to engage — to interlock with, to mesh. *Example*: Because the large gear is made _____ the small gear and turn it, the large gear is able to send power in a new direction.

The large gear is engaging the small gear.

4. to swivel — to turn freely while being joined together. *Example*: The ram is made _____ to the right or left.

5. to traverse   to go back and forth over or upon. *Example:* The table is made _____ back and forth and move the workpiece with it.

6. dovetail   a triangle shaped part that fits into a slot of the same shape; the two fit tightly together. *Example:* On some mills the knee has a _____ slot which engages the dovetail ridge on the column.

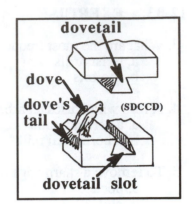

12-05: MACHINE NOMENCLATURE

Parts of the Vertical Mill

Below is a picture of a typical ram-type vertical mill. All of the main parts are shown along with some of the manual controls for the movement of these parts. Different machines will have parts, controls, and other features located in different places.

Directions: **Study this picture as you listen to the pronunciation of the parts and controls. Then do the EXERCISE below.**

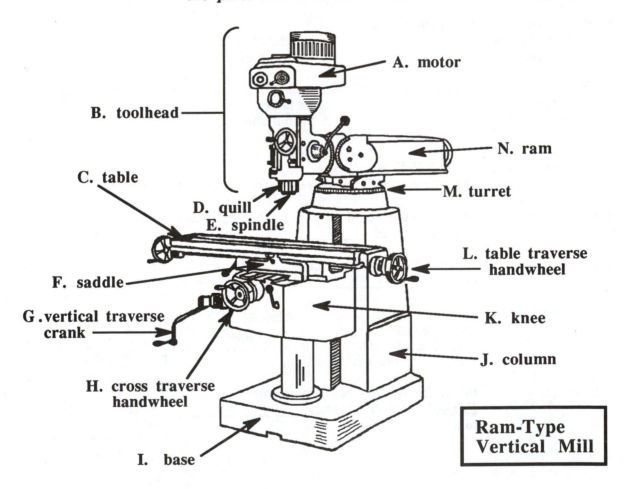

12-06: EXERCISE

Directions: Continue listening to the tape as you look at the picture on the previous page; then write your answers here.

1. _____   3. _____   5. _____

2. _____   4. _____   6. _____

12-07: READING

Directions: Read the following paragraphs.

The Main Parts and their Functions

Learning the names and functions of the main parts will help you to understand the machines and the operations they can perform. Study each of these parts and their purposes.

1. Base and Column:

The base and column are a single piece made of cast iron, and together they support and position the other major components of the vertical mill. The base and column are like the trunk of a tree that holds up the branches. The knee, saddle, table, and ram are attached to the column. The ram supports the toolhead. A **dovetail ridge** has been machined on the face of the column; the knee has a **dovetail slot** which engages the dovetail ridge; the knee slides up and down on the column. The dovetail ridge is also called the **ways**.

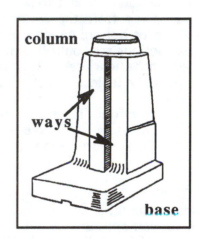

> **Function:** The purpose of the base and column is:
> a. to be a framework for attaching other parts of the mill;
> b. to support the weight of those other parts.

2. Knee:

The dovetail slot guides the vertical movements of the knee. The up and down movements of the knee are called **vertical feed** or **vertical traverse**. These vertical motions of the knee are controlled by the vertical traverse crank. This crank is turned by hand, so it is a **manual control**. For each manual control on the mill, there is usually an **automatic control**. The knee has a slide on its top which allows the saddle to slide in and out.

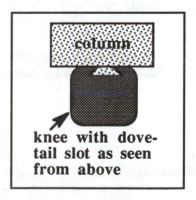

knee with dove-
tail slot as seen
from above

> **Function:** The purpose of the knee is:
> a. to give vertical movement to the workpiece;
> b. to support the saddle and the table.

12-08: EXERCISE   Directions: Circle "T" for "True" or "F" for "False."

T F 1. The base and column on the mill are a single piece of cast iron.

T F 2. "The ways" and "the dovetail ridge" on the column are the same thing.

T F 3. Most mills have automatic controls, but very few have manual controls.

12-08:   EXERCISE   (Continued)

T   F   4. The movement of the knee up and down the ways is called horizontal traverse.

T   F   5. One purpose of the base and column is to support the weight of other mill parts.

T   F   6. The vertical traverse crank is turned in order to feed the work automatically.

12-09:   READING
Directions: *Read the following paragraphs.*

More Main Parts and their Functions

Study more of these vertical mill parts and their purposes.

3. Saddle: The saddle engages the slide on the top of the knee. The saddle is able to be moved in, toward the column, or out, away from the column. The movements of the saddle in or out are called **cross feed** or **cross traverse**. These motions are controlled manually by a **cross traverse handle** or **handwheel**. The saddle has a **slide** on top of it to support the table and allow it to move right or left.

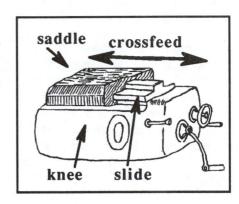

> **Function:** The purpose of the saddle is:
> a. to allow cross traverse movement;
> b. to support the table.

4. Table: The table engages the slide on the top of the saddle. The table can be moved to the right or to the left. The movements are called **longitudinal feed** or **table traverse**. The motions to the right or the left are controlled by the **table traverse handle** or **handwheel**. The work-piece or a work-holding device is secured to the table.

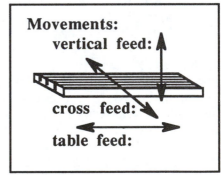

> **Function:** The purpose of the table is:
> a. to provide a surface to which the workpiece or a work-holding devise can be tightly secured;
> b. to allow table traverse movement.

12-10:   EXERCISE
Directions: Circle "T" for "True" or "F" for "False."

T   F   1. The saddle sits on top of the knee, but it has no motion of its own.

T   F   2. "Cross traverse" and "cross feed" are two names for the same motion.

T   F   3. The table provides a surface to which the work can be attached.

T   F   4. Vertical feed, cross feed, and table feed all move along the same axis.

T   F   5. The table sits on top of the saddle and moves from left to right or right to left.

T   F   6. Not all mills provide vertical feed for milling a workpiece.

More Mill Parts

5. Turret and Ram: At the top of the column there is a **turret** which allows the whole toolhead to swivel in the horizontal plane a certain number of degrees to the right and the left; there is a dovetail ridge running horizontally across the top of the turret. The **ram** engages that dovetail and can slide in over the work or back away from it. The turret and ram supports the weight of the toolhead.

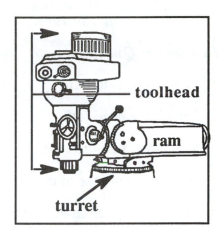

> **Function:** The purpose of the turret and ram is:
> a. to position the toolhead over the work by swiveling to the right or left, or by sliding in over the work;
> b. to support the weight of the toolhead.

6. Toolhead: The toolhead is attached to the end of the ram. In it are located the motor which is controlled by a switch (Forward, Off, Reverse). The motor powers the rotating spindle. The spindle is located at the end of the toolhead inside a quill which is a non-rotating steel tube. The quill supplies support and up-and-down motion for drilling holes.

The toolhead can also swivel at the end of the ram; the swivel is in the vertical plane like an up-and down nodding of a human head.

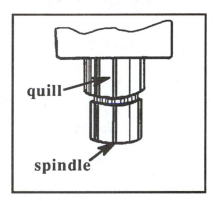

> **Function:** The purpose of the toolhead is:
> a. to house the motor and the belt or gear drives that power the spindle;
> b. to hold the spindle with its various cutters;
> c. to position the spindle toward the work.

T   F   1. The toolhead can swivel in the vertical plane as it "nods" at the end of the ram.

T   F   2. The quill rotates at the same speed as the spindle.

T   F   3. The motor is located at the end of the ram.

T   F   4. Switches or buttons that say "ON, OFF, and REVERSE" control the motor.

T   F   5. The turret allows the ram and toolhead to swivel in the horizontal plane.

T   F   6. The ram can swivel but it cannot slide in and out over the work.

T   F   7. On some mills the spindle is driven by gears; on others it is driven by belts.

12-13: "SHOP TALK"   *QUESTION & ANSWERS*

Directions:
Look at the questions below. Listen to three possible answers on the tape. Then circle the letter of the correct answer.

Instructor Question: *Student Answers:*

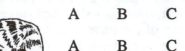

1. What's the saddle for?   A   B   C

2. What's the knee for?   A   B   C

3. What's the purpose of the base and column?   A   B   C

4. What's the purpose of the table?   A   B   C

5. What's the function of the ram?   A   B   C

6. What's the function of the toolhead?   A   B   C

12-14: "SHOP TALK"   *ANSWER & QUESTIONS*

Directions:
Look at each answer below. Listen to three possible questions on the tape. Then circle the letter of the question that asks for that answer.

Student Questions: *Instructor Answer*

A   B   C   1. Cross feed or cross traverse.

A   B   C   2. On the tool head.

A   B   C   3. Longitudinal feed or table traverse.

A   B   C   4. The manual controls.

A   B   C   5. Vertical feed or vertical traverse.

A   B   C   6. There are six.

A   B   C   7. It's controlled by a "Forward, Off, Reverse" switch.

A   B   C   8. By the sliding of the ram and the swiveling of the turret.

A   B   C   9. A dovetail ridge on the face of the column.

A   B   C   10. Because it holds the spindle, the quill and the cutter.

12-15: MACHINE NOMENCLATURE

The Main Parts of the Vertical Mill

<u>Directions</u>: *Continue learning the names of the parts and the controls. Look at the picture. Below, write the letters next to the names.*

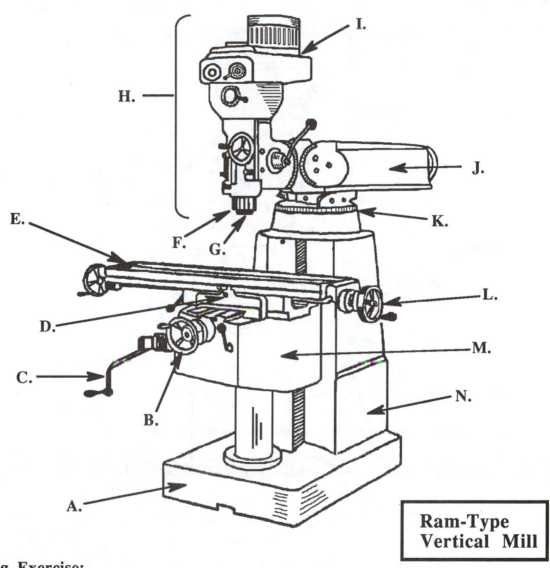

Ram-Type
Vertical Mill

Matching Exercise:

____ 1. quill	____ 6. ram	____ 10. vertical traverse crank
____ 2. base	____ 7. knee	____ 11. toolhead
____ 3. saddle	____ 8. table	____ 12. cross traverse handwheel
____ 4. column	____ 9. spindle	____ 13. table traverse handwheel
____ 5. motor		____ 14. turret

The Main Parts of the Vertical Mill

Directions: *Test yourself on the names of the parts and the controls. Look at the picture. Listen to the names on the tape. Write the letter of what you hear next to the number.*

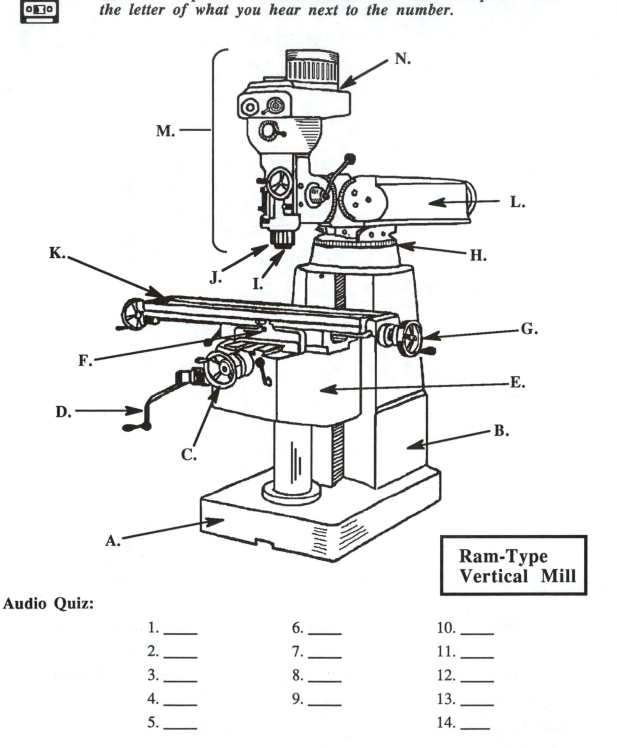

Ram-Type
Vertical Mill

Audio Quiz:

1. ____   6. ____   10. ____
2. ____   7. ____   11. ____
3. ____   8. ____   12. ____
4. ____   9. ____   13. ____
5. ____   14. ___

12-17: VOCABULARY LIST C

Directions: **The following vocabulary words are taken from the "Glossary" found in the text, Machine Tool Practices; look up each word in that glossary and fill in the blanks.**

1. to hog   to remove _____ amounts of material from a _____ with deep heavy cuts.

2. keyseat   an axially located rectangular _____ in a _____ or hub.
(or keyway)

3. periphery   the _____ or _____ boundary of a surface or body.

                                                                 (SDCCD)

4. T-nut   a threaded nut in a T shape that is designed to fit into the _____ on a machine tool _____.

5. T-slot   the _____ in a machine tool table, shaped like a _____, and used to hold T-nuts and studs for various _____ setups.

T-slots in the milling table

T-slot ⟶

(SDCCD)

12-18: EXERCISE
Directions: **Fill-in each blank below with a word from the glossary.**

1. The teeth are located on the end and on the _____ of the cutter.

2. To help with setups, the milling table has _____ s which will take _____ s .

3. To get down to a finishing cut, the machinist is able _____ away a large amount of material with a roughing end mill.

4. Both the gear and the shaft must have a _____ cut into their sides.

12-19: READING
Directions: Read the following paragraphs.

Some Cutters for the Vertical Mill

Cutters used on a vertical mill have shanks. The shanks can be tapered or straight. The shank of a milling cutter must be inserted into the spindle of the mill. The inside of the spindle usually has a taper, so the shank must be tapered, or a tapered tool-holder used. There are several different kinds and sizes of tapers found on mills and on cutters. When a tapered shank or tapered tool-holder is inserted into the spindle, an outside taper is mated with an inside taper; they will hold together very tightly. The two parts must be separated by force.

Learn the names of some common milling cutters and the milling tasks for which each can be used.

1. End Mills
The end mill gets its name from the fact that it primarily uses teeth at its end to do cutting jobs, but the flutes on the periphery are also used for cutting; occasionally the end mill is used for plunge cutting, like a drill. It is the most frequently used cutter for the vertical mill.

a. Classifications of end mills:

(1) <u>by number of flutes</u>: There are end mills with one, two, three, four or more flutes.

(a) the two flute end mill can take a large chip by cutting deeper into the material.

(SDCCD)

two-flute, double-end end mill

(b) the four flute end mill is stronger than the others and has a better finishing cut. This also allows for a faster feed rate.

(2) <u>by direction of cutting</u>:

(a) left-handed cutters: They make their cuts by turning in a clockwise direction.

left-hand cutting   **right-hand cutting**

clockwise = in the direction in which the hands of a clock move

(b) right-handed cutters: They make their cuts by turning in a counterclockwise direction.

counterclockwise = in the direction opposite of the way the hands of a clock move

(3) <u>by the number of ends</u>:
(a) single end mills have a shank at one end and cutting teeth at the other.
(b) double end mills have cutting teeth at both ends. Only one end is used, while the other end fits inside a collet or other holding device; with two ends, the second end will be ready for use when the first end becomes dull.

b. Various kinds of end mills:

(1) <u>straight</u> or <u>helical flutes</u>: the flutes can be straight or helical with a variety of helix angles from "slow" (12°) to "regular" (30°) to "fast" (45°).

(2) with corner radius: the corner radius prolongs tool life by preventing corner chipping; produces a fillet when cutting pockets.

(3) <u>tapered</u>: used for making precise tapered or angled surfaces.

(4) <u>roughing end mill</u> (or hogging end mill): has wavy teeth on the periphery of the cutter. Used for removing large amounts of material from the workpiece.

**roughing end mill
or hogging end mill**

(5) <u>ball-end mill</u>: can be used for machining round-bottom grooves

two-flute ball-end end mill

(6) <u>single-angle milling cutter</u>: can be used for machining dovetails.

(SDCCD)

single-angle milling cutter

(7) <u>T-slot cutters</u>:
This cutter is used to cut the arms of the "T" after a vertical slot has been cut. In the first picture a slot has been cut in the workpiece, the vertical stroke of the "T"; in the second picture, the T-slot cutter is used to cut the two arms of the "T."

(SDCCD)

T-slot milling cutter

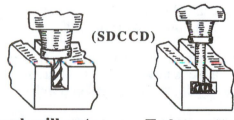

(SDCCD)

**end mill cuts
the vertical
stroke of
the "T"**

**T-slot cutter
cuts the arms
of the "T"**

(8) <u>Woodruff keyseat cutters</u>: to cut Woodruff keyseats. In the picture this key and keyseat are designed to hold a gear in place on the end of a shaft.

Woodruff key

shaft →

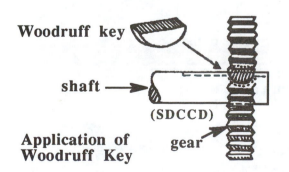

(SDCCD)

**Application of
Woodruff Key**   **gear**

**WOODRUFF
KEYSEAT
CUTTER**

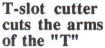

(SDCCD)

155

(9) <u>flycutters</u>: The flycutter has one or more lathe toolbits held in place in a special holder by one or more set screws. Such a cutter can be used to remove material from a large surface area (a face cut); the flycutter can also be used for boring operations.

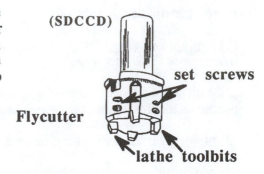

(SDCCD)

set screws

Flycutter

lathe toolbits

(10) <u>shell end mills</u>: this cutter looks like a shell with the curl of its blades. It is locked in place on an adapter which fits into the taper of the spindle. This mill is also used for face milling.

Shell end mill
(SDCCD)

Adapter

12-20:   EXERCISE
Directions: *Read the questions.   Circle the **best** answer.*

1. A cutter which makes its cuts by turning in a counterclockwise direction is called:
    a. a right-handed cutter   c. a counter cutter
    b. a left-handed cutter   d. none of these

2. A cutter useful for removing large amounts of material from a workpiece is called:
    a. a ball-end mill   c. a single-angle milling cutter
    b. a double-end end mill   d. a hogging end mill

3. What features do you find on a flycutter?
    a. set screws and T-slot cutters   c. set screws and lathe toolbits
    b. adapters and shell end mills   d. a ball-end mill and a split collet

4. What cutter is used to cut dovetail slots?
    a. a T-slot cutter   c. a single-angle milling cutter
    b. a dovetail cutter   d. tapered end mills

5. The tool in the picture at the right:
    a. is held in place with a set screw. c. has two flutes.
    b. is used one end at a time.   d. all of these.

(SDCCD)

6. A helical flute is called "fast," if it has a helix angle of:
    a. 30°   b. 45°   c. 90°   d. none of these

7. A benefit of using a cutter with a corner radius is:
    a. prevention of corner chipping   c. both a and b
    b. prolonged tool life   d. neither a nor b

8. The tool on the right is used to make:
    a. dovetail slots   c. T-slots
    b. Woodruff keyseats   d. pockets

(SDCCD)

12-21: READING
Directions: *Read the following paragraphs.*

More about Cutters for the Vertical Mill

2. Ways to hold cutters:
There are several ways to hold the shank-type milling cutters. They must all finally fit into the vertical mill's spindle.

a. Collets:
　　　(1) <u>solid collets</u>: The collet has a shank which fits into the spindle. It has a hole in the bottom into which fits the shank of the cutter. There are set screws that can be turned snugly against the flat spot on the shank of the cutter. There are a variety of sizes in solid collets to accept a variety of cutter shanks.

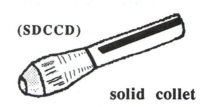

(SDCCD)

solid collet

　　　(2) <u>split collets</u>: This collet has slits in the sides of the collet which are squeezed together around the cutter shank when the collet is drawn into the spindle taper.

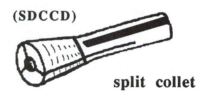

(SDCCD)

split collet

b. Quick change systems:
These systems are set up ahead of use with different tools mounted in individual toolholders; these toolholders are then quickly mounted or removed from a master toolholder. (*Look at page 526 in the textbook for pictures and page 527 for further details.*)

12-22: PRACTICE "SELF-TEST" QUESTIONS:
Directions: *Write complete answers; then check the answers on page 775 in the textbook (Section J, Unit 2).*

Q. 5: To remove a considerable amount of material what kind of end mill is used?

Q 8: How are straight shank tools held in the machine spindle?

Q. 10: Why are quick-change toolholders used?

To communicate successfully in the shop, a worker should know how to make requests and to understand them. A request is asking someone for something. Requests are more likely to be successful, if they are made politely.

Directions:
Read Le's statements below. Listen to three possible requests on the tape. Then circle the letter which goes best with the statement.

1. I'm going to cut some T-slots.   A   B   C

2. I'm going to attach this gear to that shaft.   A   B   C

3. I want to take a quarter inch off this whole steel surface.   A   B   C

4. I'm going to cut a dovetail in this steel block.   A   B   C

5. I need to machine some grooves with a full radius.   A   B   C

6. I want to get a solid hold on this double-end, two-flute mill.   A   B   C

12-24:   MACHINE NOMENCLATURE

Vertical Milling Cutters

Below are pictures of some of the cutters for the vertical mill. Learn the names of these cutters and the others in your textbook.

Directions: *Study this picture as you listen to the pronunciation of these cutters on the tape.*

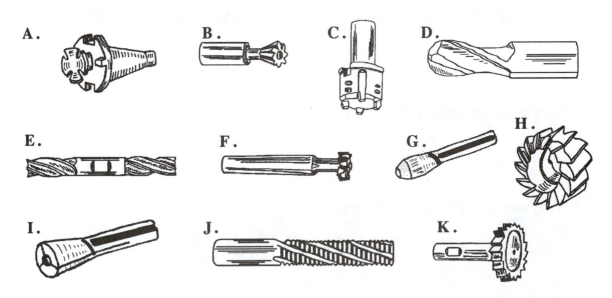

A.   B.   C.   D.

E.   F.   G.   H.

I.   J.   K.

12-25:   EXERCISE
Directions: *Look at the pictures in 12-24. Write the letters next to the names.*

____ 1. single angle cutter   ____ 5. roughing end mill   ____ 9. Woodruff keyseat cutter

____ 2. shell end mill   ____ 6. flycutter   ____ 10. T-slot cutter

____ 3. solid collet   ____ 7. split collet   ____ 11. two-flute, double end mill

____ 4. two-flute ball-end   ____ 8. adapter   ____ 12. hogging end mill
      end mill

12-26:   VOCABULARY LIST D
Directions: **Study the vocabulary and fill in the blanks with the missing words.**

1. to make sure — to do something carefully; to take care that something is done. *Example*: He has _____ that the workpiece is securely fastened.

2. to expose — to be out in the open air, to leave uncovered. *Example*: Be sure not _____ any gears or belts when they are running. Cover them with a guard.

3. contaminated — poisoned; having dirt in it. *Example*: Cutting fluid becomes _____with chips, dirt, and small particles of metal.

4. cut — an opening in the skin made by something sharp. *Example*: Walter washed his hands and covered the _____with a cloth.

      sore

      cut

5. sore — an opening in the skin which is usually infected and painful. *Example*: Be careful not to get dirt or cutting fluid in cuts and _____s

6. tired — the condition of someone who has used up all strength and energy and who needs to rest. *Example*: I'm too _____ to work now.

7. sleepy — the condition of someone who is about to go to sleep. *Example*: It is very dangerous to operate an automobile or a milling machine when you are _____.

8. upset — the condition of someone who is angry or has other strong emotions. *Example*: It is also dangerous to drive or operate machinery if you are worried or _____, so that you don't keep your mind on your work.

9. coordination — muscles working together to produce accurate movements. *Example*: When you drink alcohol, you lose your _____.

10. to take precaution — to be careful. *Example*: When you operate a mill, make sure _____ by securing the workpiece and the cutting tool.

12-27: READING
Directions: Read the following paragraphs.

Safety and the Vertical Milling Machine

The vertical mill is a very powerful tool. You have seen how rapidly the spindle turns and how the cutters can remove large amounts of material from a workpiece. It is important to learn and remember the rules of safety for operating this important machine.

Hazards	Safety Rules
1. Workpieces and/or cutters can break and be thrown through the air.	Make sure that: • you **wear your safety glasses** at all times in the shop; • you watch out for coworkers and their machines; • the cutting speeds and feed rates are correct.
2. Workpieces and/or cutters can come loose.	Make sure that: • the workpiece is securely fastened to the table; • the cutter is held securely in the spindle.
3. Spindles and cutters rotate very rapidly.	Make sure that nothing gets caught in the moving parts: • long hair should be pinned up or held in a hair net; • jewelry should not be worn; • wear short-sleeve shirts or roll long sleeves to above the elbows.
4. The cutters are very sharp and heavy.	Make sure to use a shop rag to carry any cutters from one place to another.
5. The machinist often needs to measure the workpiece.	Make sure the motor is turned off before you measure the workpiece.
6. Gears, belts and other moving parts can be exposed.	Make sure all safety guards are in place on the machine.
7. Milling machine chips are hot, sharp, and contaminated with cutting fluid.	Make sure to: • not use your hands to brush off chips; • not use compressed air to clean up chips; • use a brush or vacuum cleaner to get up chips; • wash your hands after work and keep cuts covered.
8. You may work in a safe manner, but co-workers may not.	Make sure to report unsafe practices or conditions that your coworker(s) may have.
9. You may be using a machine with controls that you are not completely familiar with.	Make sure to: • ask questions when you are not sure; • know what will happen when you turn on the machine; • learn how to operate the controls of each machine; • know where the ON/OFF switch is located.

160

Hazards	**Safety Rules**
10. You may come into the shop feeling tired, sick or upset; you may be taking medicines that affect your sight or coordination.	Do not operate the machines if you are not completely alert and able to work through your shift. Tell your teacher or supervisor about your condition.

12-28: "SHOP TALK" QUESTION AND ANSWERS

Directions:
*Read each question. Listen to three possible answers. Then circle the **best** answer for each question.*

1. What is the best way to protect your eyes when using the vertical mill?   A   B   C

2. What precaution should you take when carrying a milling cutter?   A   B   C

3. What do you do when you want to measure a workpiece?   A   B   C

4. What precaution should you take when working near the mill with its moving parts, like gears, belts, and drives?   A   B   C

5. What precautions should you take if you come to work with an open cut or sore?   A   B   C

6. What should you do in this situation: A coworker is using compressed air to blow chips away from his workpiece; you also see his head nod several times; he is tired and sleepy.   A   B   C

12-29: EXERCISE

Directions: Turn to page 518 of the textbook and find the "Shop Tip" in the red box. Read the information. Then fill in the missing words below.

toolhead

ram

Shop Tip

Extreme _____ has to be exercised when the workhead needs to be _____ or tilted to cut _____ surfaces. After_____ the clamping bolts that hold the toolhead to the ram, _____ them slightly to create a light drag. Never loosen _____ the clamping bolts _____. There should be enough friction between the toolhead and the ram that the toolhead _____ only when _____ is applied to it. If the clamping bolts are loosened completely, the _____ of the heavy spindle motor will flip the toohead _____ down. This can cause serious _____ to the operator and _____ the toolhead and the machine table. This operation should be performed by ____ persons.

161

Vertical Milling Cutters

Directions: Test yourself on the names of the vertical milling cutters. Look at the pictures. Listen to the names on the tape. Write the letter of what you hear next to the number.

A.　　　B.　　　C.

D.　　　E.

F.　　　G.　　　H.

I.　　　J.　　　K.

Audio Quiz:

1. ____   4. ____   7. ____   10. ____
2. ____   5. ____   8. ____   11. ____
3. ____   6. ____   9. ____   12. ____

Directions: Look at the pictures. Listen to the uses for the cutters. Write the letters next to the numbers.

Cutter Uses Quiz:

13. ____   16. ____   19. ____   22. ____
14. ____   17. ____   20. ____   23. ____
15. ____   18. ____   21. ____

Unit 13: *HORIZONTAL MILLING MACHINES*

ASSIGNMENT: **Read, study and complete pages 163 to 172 of this book. Then read pages 549 to 595 in the textbook,** *Machine Tool Practices,* *SECTION K, Horizontal Milling Machines.*

OBJECTIVES for this unit:
You should be able to:
1. Identify and say correctly the important parts of the horizontal milling machine.
2. Identify and say the names of the milling machine controls.
3. Identify and pronounce correctly the names of some common arbor-driven cutters.

13-01: VOCABULARY LIST A
Directions: Study the vocabulary. Write the missing words in the blank spaces.

1. support

a part of a machine which carries the weight of other parts and holds them in place. *Example:* Each arbor _____ holds the arbor and aligns it with the spindle. Inside each arbor _____ are bearings for holding the arbor while leaving it free to turn.

(SDCCD)

arbor supports

2. housing

a frame or box for holding some mechanical parts. *Example:* A motor is usually placed inside a _____ which will protect the operator from being injured by the moving parts.

3. interface

The place where two things meet or touch. *Example:* The picture shows the tool-chip _____; that is where the tool cuts into the work and forms a chip.

(SDCCD)

tool-chip interface

4. reservoir

a place where a liquid is collected and stored. *Example:* The base of the milling machine can have a _____ in which cutting fluid or coolant can collect.

5. manual

done by the use of the hand. *Example:* Gloria prefers to use the _____ controls, instead of the automatic ones, when she uses the mill

6. rigidity

condition of being unmoving and firmly fixed. *Example:* It is good to use a mill with _____ in its major parts; the results will be more accurate.

7. dial

the face of a meter, gage, or indicator on which a pointer indicates an amount or direction or speed. *Example:* An operator can turn the speed _____ to choose the speed at which he/she will run the mill.

8. range

the full set of numbers which show how much something moves or operates. *Example:* Using the horizontal mill, the machinist can choose the _____ of speeds at which to run the machine.

163

13-02: READING
Directions: Read the following paragraphs.

Parts of the Horizontal Mill

Like the vertical milling machine, the **horizontal milling machine** used in many machine tool classrooms is of the **knee and column type**, in which a **knee** moves up and down a **dovetail slide** on the face of a **column**; this motion raises and lowers a **table** into a cutter which is mounted on a rotating **arbor**.

One end of the arbor fits into a **spindle** which drives the arbor; the other end of the arbor is held in **arbor supports**. Each arbor support has **bearings** which hold the arbor but allow it to freely rotate; the arbor supports slide onto an **overarm**, which extends out over the table. The overarm attaches to the top of the column by fitting onto a **dovetail slide**.

If the knee and column milling machine has a **table swivel** inside a **table swivel housing**, the table can be swiveled in the horizontal plane up to 45° to the left or right. If it has this swiveling table, the machine is called a **universal** milling machine, because it can do more jobs; without the swiveling table, it is called a **plain** milling machine.

The table has **T-slots** to which work or work-holding devices can be attached. A **coolant hose** supplies coolant to the tool/chip interface. An open hollow in the top of the base is used as **chip pan** to catch falling chips and to be a **reservoir** for coolant.

13-03: NOMENCLATURE
Directions: Listen to the names of each part; practice pronouncing each word; rewind, and write the names as you listen again.

C. overarm   D. dovetail slide
B. spindle
A. column   E. arbor supports
D. dovetail slide   F. bearings
M. coolant hose   G. table
L. saddle   H. table swivel housing
K. base   I. knee
    J. chip pan

Universal Horizontal Milling Machine   (SDCCD)

A. _____
B. _____
C. _____
D. _____
E. _____
F. _____
G. _____
H. _____
I. _____
J. _____
K. _____
L. _____
M. _____

164

13-04: EXERCISE

Directions: *For each statement, circle T for "true" or F for "false."*

T   F   1. A plain knee and column milling machine does not have a table swivel.

T   F   2. The spindle and the arbor supports hold the rotating arbor.

T   F   3. T-slots are cut into the base in order to hold the workpiece.

T   F   4. The overarm is attached to the column by sliding onto a dovetail.

T   F   5. The coolant hose is used to spray coolant on the motor.

T   F   6. The table swivel is inside the table swivel housing.

T   F   7. A 45° table movement in the horizontal plane is only possible using a plain horizontal mill.

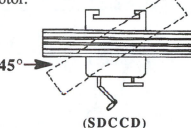

(SDCCD)

13-05: READING

Directions: *Read the following paragraphs.*

Controls of the Horizontal Mill

The horizontal mill is equipped with controls for feeding the work into the cutters: There are automatic power feeds for the table (**power table feed lever**), the saddle (**power cross feed lever**) and the knee (**power vertical feed lever**). There are also manual feeds for the table (**table feed handwheel**), for the saddle (**cross feed handwheel**) and the knee (**vertical feed hand crank**). The table, the saddle, and the knee can each be locked in place to give greater rigidity during machining; the picture shows one of these controls, the **knee clamp**.

The spindle speed is controlled by the **speed change dial** located on the side of the column; this control allows the operator to set the number of RPMs at which the spindle is turning. Several RPM speeds are available in either a high range or a low range; the range is set by using a **high/ low speed range lever**.

Horizontal mills will always have a **start/stop switch** for turning the machine on and off.

13-06: CONVERSATION

  Directions: *Lucy Garcia and Omar Harris have come to their teacher, Al Lopez, for advice on memorizing the names of the controls.*

Lucy: Mr. Lopez, we're trying to memorize the names of the controls. Can you help us?

Al: Lucy, it sometimes helps if you can put things together in groups. With the controls on this mill, you have two main groups: the manual and the automatic. Then you know they're contolling the movements of three basic parts: the table, the saddle, and the knee. When you put those together you have three automatic power controls and three manual controls. All the automatic controls are levers.

Omar: So, let's see: Automatic controls are power vertical feed lever, power cross feed lever, and power vertical feed lever. The manual controls will be two handwheels-- the table feed handwheel and the the cross feed handwheel--and one hand crank-- the vertical feed hand crank.

13-07: NOMENCLATURE

 Directions: Listen to the names of each part; look at the pictures of the controls; practice pronouncing each word; rewind, and write the names as you listen again.

Controls of the Horizontal Mill

N. _____

_____

O. _____

_____

P. _____

_____

Q. _____

_____

R. _____

S. _____

_____

T. _____

_____

U. _____

_____

V. _____

_____

W. _____

_____

Horizontal
Milling Machine   (SDCCD)

__Horizontal Mill Controls:__
N. start/stop control
O. high/low speed range lever
P. speed change dial (RPM)
Q. table feed handwheel
R. knee clamp
S. power table feed lever
T. power cross feed lever

U. cross feed handwheel
V. vertical feed hand crank
W. power vertical feed lever

13-08: EXERCISE
__Directions__: For each statement, circle T for "true" or F for "false."

T F 1. The knee is the only part of the horizontal mill that can be locked.

T F 2. Locking some parts during machining gives the mill greater rigidity.

T F 3. Table, saddle, and knee movements all have automatic and manual controls.

T F 4. Automated movements of the saddle are called vertical feed.

T F 5. The speed change dial and the high/low speed range lever are used to set RPMs.

13-09: VOCABULARY LIST B
Directions: *Study the vocabulary. Write the missing words in the blank spaces.*

1. spacing collar   a horizontal milling accessory which is a short steel tube which slides onto the arbor as a means of spacing the cutter on the arbor. *Example:* Be sure that each _____ on the arbor is clean and free of scratches.

spacing collars
(SDCCD)

2. arrangement   the way a group of things are laid out for use or appearance. *Example:* The _____ of the teeth on the periphery of the cutter is circular.

3. tendency   a leaning toward acting in a certain way *Example:* Some cutters have a _____ to chatter, especially if the work or the tools are not very rigid.

staggered teeth

4. to stagger   to arrange things along a line with things going from one side to the other: right side, left side, right side, etc. *Example:* The teeth on some milling cutters are made _____ from one side to the other.

(SDCCD)

5. alternate   placed first on one side then on the other; every other one. *Example:* Joe and Jaime share a job. They work on _____ days of a 6-day week: Joe works Mon., Wed., Fri., and Carlos works Tues., Thurs., and Sat.

6. to bind   to become caught or stuck in a narrow place, so that further motion is not possible. *Example:* Without suitable clearance the saw is going _____ in the slot it is cutting.

7. to shatter   For something hard to break into sharp, jagged pieces because of pressure or a blow. *Example:* If the feed is too deep, the speed too high, or the blade is too rigid, a saw is going _____ when it hits the work.

8. to slit   a long, thin, straight cut made in any material. *Example:* He used a letter opener _____ the envelope and take out his pay check.

9. to vary   for each thing in a group to be different in some respect from each other. *Example:* The angles on an angular milling cutter are able _____ from 45° to 60°.

10. to square   to make sure that all the right angles on a rectangular workpiece or an arrangement are as close to 90° as possible. *Example:* Sometimes it is necessary _____ the workpiece before beginning the major cuts.

Squaring the top of a workpiece with a face milling cutter

Arbor-Driven Cutters and the Horizontal Mill

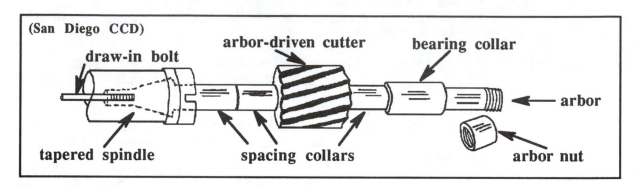

The vertical mill uses **shank-type** or **adapter-held cutters**; adapters can also be used to hold some tapered shank-type tools in the horizontal mill. However, the horizontal mill mainly uses **arbor-driven cutters**; such a cutter is round like a wheel with a hole in the center through which an arbor is passed. The cutter itself is positioned along the arbor by **spacing collars**, which come in a variety of lengths, and by **bearing collars** which fit inside the arbor support bearings. The surfaces of these collars must be protected from scratches and from dirt to insure that they are accurate.

Both the spindle and the tapered end of the arbor have a taper of about 16 1/2° as measured from the axis. The **standard national milling machine taper** on the arbor and in the **spindle** is 3 1/2 inches per foot; this taper is for alignment rather than for holding. To hold the tapered end of the arbor in place, a **draw-in bolt** passes through the middle of the spindle and screws into the end of the arbor; the arbor can be pulled into the spindle. An **arbor nut** is used to lock the cutters on the arbor.

What are some of the cutters that can be mounted on an arbor?

1. Plain Milling Cutters: These cutters are used to mill plain surfaces that are more narrow than the cutter. These cutters can be either straight or helical.

The **straight plain milling cutter** has the edges of its cutting teeth parallel with the axis of the arbor; in this arrangement the tooth is cutting all along its edge at the same time; these cutters have a tendency to chatter. Straight teeth are only found on plain milling cutters less than 3/4 in. wide.

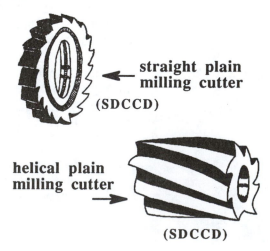

The **helical plain milling cutter** has the teeth curling around the length of the cutter in a helix; the helical cutting edges allow each tooth to make a gradual cut; this results in a smoother cut with less chatter. There are no cutting teeth on the sides of these cutters.

168

13-11: EXERCISE
Directions: For each statement, circle T for "true" or F for "false."

T F 1. Straight teeth are found only on plain cutters that are less than 3/4 inch wide.

T F 2. The standard national milling machine taper is 3/32 inches per foot.

T F 3. The standard national milling machine taper is 3 1/2 inches per foot.

T F 4. The milling machine taper will hold the arbor in place, but will not align it correctly.

T F 5. A draw-in bolt will hold the arbor in place.

T F 6. A bearing collar fits into the arbor support bearings.

T F 7. A milling cutter can be positioned on the arbor by the use of space collars.

T F 8. Straight plain milling cutters make smoother cuts and chatter less than helical ones.

13-12: READING
Directions: Read the following paragraphs.

More Arbor-Driven Cutters

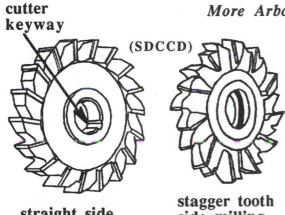

straight side milling cutter

stagger tooth side milling cutter

2. Side Milling Cutters: The side milling cutter has teeth on its periphery and on its side or sides. Side milling cutters are used to cut steps, grooves, shoulders, and deep slots. The widths of these cutters range from 3/16 to 1 inch. There are both **straight side milling cutters** and **stagger tooth side milling cutters**. The cutter is called "stagger tooth" because it has one tooth on the right and the next tooth on the left, and so on around the periphery. For deep grooves the stagger tooth side milling cutter is better, because it gives a smoother cut with more clearance provided by its alternate right-hand and left-hand helical teeth.

Above left, the picure of the straight side milling cutter shows a **cutter keyway** in the circumference of the central hole. This picture at the lower left shows a machinist placing a stagger tooth cutter onto the arbor: A **key** has been placed into the **arbor keyway** and the spacing collars have been aligned to fit onto the key; as the machinist pushes the cutter onto the arbor, the keyway on the cutter is lined up to slide over the key. This is how a cutter is held securely.

Notice also how the machinist holds the cutter with a pair of shop rags to avoid cutting the hands.

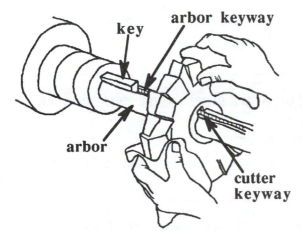

169

3. Plain Metal Slitting Saw: The **plain metal slitting saw** is used for cutting slots and for cutting off lengths of rough stock for use with various machine tools. The sides of this saw are curved slightly inward toward the central hole to provide some clearance which will lessen the chance of the saw binding in the slot. The width of these saws ranges from 1/32 to 5/16 in.

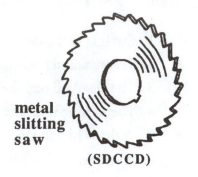

metal slitting saw

(SDCCD)

When mounting the saw on the arbor, the key is left out, in order to allow the arbor to turn freely, if the blade becomes bound in the work. Otherwise the blade might shatter.

The plain saw is used for light cuts. For deeper cuts there are the **side tooth metal slitting saw** (Figure K-35 in the textbook) and the **stagger tooth metal slitting saw** (Figure K-36).

single angle cutter

(SDCCD)

4. Angular Milling Cutters: Like similar cutters used on the vertical mill, the angular milling cutter is used to cut notches and dovetails and chamfer edges. This picture shows a single angle cutter; the angle of the cutting teeth of such cutters can vary from 45 to 60 degrees. In the textbook, Figure K-39 shows a double-angle milling cutter.

5. Concave and Convex Milling Cutters: Some special milling cutters can be used to cut special shapes. A **convex cutter** will cut concave grooves, and a **concave cutter** will cut convex ridges. A third type, the corner-rounding milling cutter, is shown in Figure K-42; it can be used to round corners.

(SDCCD)

convex   concave

13-13:   CONVERSATION

Directions: *The students are reviewing arbor-driven milling cutters with Al. Read the questions and write your answers. Listen to the tape for the answers given by Le, Lucy, and Omar.*

Al: Good Morning! Can anyone tell me this: What milling cutter is good for cutting deep slots and has good clearance and a smooth cut?

Lucy: _____

Al: That's right, Lucy! Now someone tell me what a metal slitting saw is used for.

Omar: _____

Al: You're right, Omar. Now let me ask: How can I cut a groove with a full radius?

Le: _____

Al: Good answer, Le. You all are doing well. Keep studying and practicing.

170

13-14: READING
Directions: Read the following paragraphs.

Face Milling Cutters

The **face milling cutter** is typically <u>not</u> mounted on an arbor, but is often used in a horizontal mill, as well as in a vertical mill. It is mounted directly on the spindle nose of the horizontal mill or on a centering plug which fits into the spindle; see the textbook: Figures K-106 and K-107. Face milling cutters have teeth which are inserted into the body of the cutter. The **inserts** can be made of high speed steel, or tipped with carbide; the teeth can cut on the periphery and on the face of the cutter. A face milling cutter of less than 6 inches is sometimes called a **shell end mill**; it is usually fitted onto an **adapter** which is inserted into the spindle.

The face milling cutter is used to square a rectangular workpiece by machining flat surfaces and making sure the edge angles approximate 90° as closely as possible. These cutters are also used for cutting shoulders.

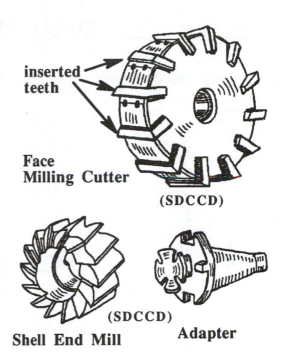

inserted teeth

Face Milling Cutter

(SDCCD)

(SDCCD)

Shell End Mill

Adapter

13-15: EXERCISE
Directions: For each statement, circle T for "true" or F for "false."

T F 1. A face milling cutter can be mounted on a horizontal or a vertical mill.

T F 2. A face milling cutter is usually thrown away when the teeth become dull and worn.

T F 3. A face milling cutter less than 6 inches in diameter can be called a shell end mill.

T F 4. A shell end mill can be used for cutting a shoulder on a workpiece.

T F 5. A face milling cutter may have inserted teeth made of high speed steel or with carbide tips.

13-16: MACHINE NOMENCLATURE   *AUDIO QUIZ*

Directions: Test yourself on the names of the parts of the horizontal milling machine, the milling arbor, and the milling cutters. (You may want to review before your take the test.) Look at the picture. Listen to the names on the tape. Write the letter of what you hear next to the number.

1. _____   5. _____
2. _____   6. _____
3. _____   7. _____
4. _____

D.

(SDCCD)

A.   B.   C.   E.   F.   G.

8. ____   25. ____
9. ____   26. ____
10. ____   27. ____
11. ____   28. ____
12. ____   29. ____
13. ____   30. ____
14. ____   31. ____
15. ____   32. ____
16. ____   33. ____
17. ____   34. ____
18. ____   35. ____
19. ____   36. ____
20. ____   37. ____
21. ____   38. ____
22. ____   39. ____
23. ____   40. ____
24. ____   41. ____

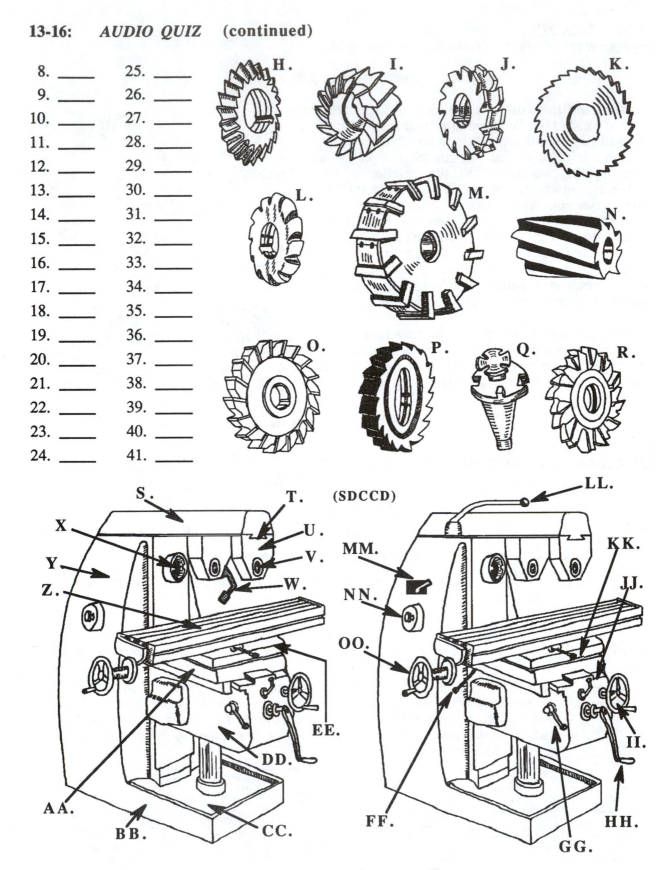

(SDCCD)

Unit 14: *GRINDING MACHINES*

ASSIGNMENT: **Read, study and complete pages 173 to 188 of this book. Then read pages 631 to 706 in the textbook,** *Machine Tool Practices,* *SECTION N, Grinding and Abrasive Machining Processes.*

OBJECTIVES for this unit:
You should be able to:
1. Identify and say correctly the important parts of three common grinding machines.
2. State the important reasons to choose a particular grinding wheel for a particular job.
3. Explain how to true, dress, and balance a grinding wheel.

(SDCCD)

14-01: VOCABULARY LIST A
Directions: *Study the vocabulary.   Write the missing words in the blank spaces.*

1. abrasive grain   a tiny, hard, solid piece of abrasive material. *Example:* Each _____ _____ in the grinding wheel can cut a tiny chip from the metal.

2. to bond   to hold pieces together with material like glue or solder. *Example:* Rubber can be used _____ abrasive grains together to make a grinding wheel.

pedestal grinder

3. swarf   the large number of tiny chips that are produced when a grinding wheel is used remove metal. *Example:* Many grinding machines have a system to collect and safely remove the _____ produced during grinding operations.

4. friable   easily fractured into tiny pieces or easily crushed into powder. *Example:* If an abrasive grain is _____, it will break with heat and pressure and produce new cutting edges.

5. economical   a way of doing something in which there is no waste of money or time. *Example:* Eduardo decided it was more _____ to make the rods on a grinder than on any other machine.

6. to fuse   to unite by melting things together, by changing them to liquid by means of heat. *Example:* The maintenance machinist repaired the machine when he was able _____ the two broken pieces.

7. alloy   a mixture of two or more metals or a metal and some other chemicals, all fused together by heat. *Example:* Each different kind of steel is an _____ made of iron and some other metals or chemicals.

8. ceramic   made of baked clay or other material from the earth. *Example:* Some _____s_ are very useful in the machine shop.

9. ferrous   containing iron. *Example:* The _____ metals include all of the steel alloys. Copper and aluminum are *non-ferrous metals.*

14-02: READING
Directions: Read the following paragraphs.

SOME FIRST IDEAS ABOUT GRINDING

In the machine trade a grinding machine does its jobs by cutting out very tiny chips from a workpiece. The chips are not made by drills, or toolbits, or cutters, but rather by the cutting edges of many small **abrasive grains** which are bonded together in a **grinding wheel** or other solid form. The large number of little chips that result from the grinding process is called **swarf**.

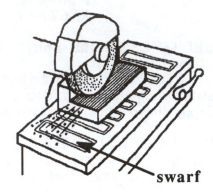
swarf

Grinding Wheel

Some abrasive grains have a property called **friability** which means that the grains will fracture as heat and pressure build up on the grinding surface; when the grains break they expose new cutting edges which continue to remove more swarf.

The cutting action of a grinder can be applied in a **variety of operations**:

(1) A typical use is to put a finish on work that has been shaped on some other machine. A grinder can remove tiny amounts (less than .001 in). from a metal surface; this can give the final degree of accuracy to a part's dimensions.

(2) the final appearance of a part may be made very smooth, by grinding.

(3) grinding may be a more economical way to remove material from work than lathes or mills are, especially if the material is very hard or has been heat-treated. If an entire part is made by grinding, this is called **abrasive machining**.

(4) another common use is the sharpening of cutting tools for drill presses, lathes and mills.

(5) there are several kinds of grinders which allow for a great variety of applications to round work or flat work, to small work or to large pieces.

14-03: EXERCISE
Directions: Match the letters on the right with the words on the left.

____ 1. grain   a. an efficient, no-waste way of doing something.

____ 2. swarf   b. a surface grinder can remove surface amounts of this size.

____ 3. friable   c. an example of this is putting the final finish on a part.

____ 4. economical   d. the cutting edges of the grains remove tiny pieces of metal.

____ 5. bonding   e. a quality by which some abrasive grains are easily fractured.

____ 6. abrasive cutting action   f. using a material that that will hold the grains together in a shape like a wheel.

____ 7. .0005 in.   g. the large number of tiny chips that result from grinding.

____ 8. grinder operation   h. a tiny, hard, solid piece of material.

14-04: READING
Directions: Read the following.

SELECTING A GRINDING WHEEL

The machinist must plan how a job will be done; this includes selecting the proper tools. For grinding operations, this means choosing the best kind of grinding wheel for a particular job. Wheels will vary in the type of **abrasive**, **grit size**, **hardness**, **structure**, and **bond**. On this page and the next one, study each of these things used to choose a grinding wheel.

1. Type of abrasive: In order to cut materials the abrasive grains must be harder than the material they are cutting; hardness will be important in choosing an abrasive.

14-05: EXERCISE

Directions: Listen to the following summaries of pages 647-648 in the textbook; fill in the blank spaces with what you hear on the tape; check in the textbook to see if you are correct.

a. _____ **aluminum** _____: This abrasive is often used on smaller shop grinders, called pedestal grinders, for sharpening tools. It works well on steels and steel _____, but not on cast _____.

b. _____ _____ **oxide**: This can be used for _____ grinding, but not for precision grinding. It works well in combination with fused aluminum oxide to make a _____ with longer life than fused aluminum oxide alone.

c. _____ **carbide**: This abrasive is harder than aluminum oxide and has sharper, more _____ crystals. It works well on cast iron, and the alloys of non-ferrous metals like aluminum and copper.

d. cubic _____ _____ (CBN) or **Borazon**: CBN is even harder than silicon carbide and works well on hardened _____ alloys like those using cobalt and nickel, as in the aircraft industry.

friable grains

e. _____: This is the hardest material known, so it is capable of cutting anything else. It occurs in nature and can now be _____. Diamond is used in preparing and renewing the surfaces of grinding wheels made of other materials; diamond is useful too in sharpening _____ cutting tools.

14-06: EXERCISE
Directions: Match the letters on the right with the words on the left.

____ 1. fused aluminum oxide   a. the hardest material known; it can be manufactured.

____ 2. ceramic Al_2O_3 b. necessary quality for one thing to cut another.

____ 3. silicon carbide   c. works well on hardened ferrous alloys.

____ 4. Borazon (CBN)   d. used frequently in the shop for sharpening small tools.

____ 5. diamond   e. works well on cast iron, aluminum, and copper.

____ 6. hardness   f. works well in combination with fused aluminum oxide.

14-07: VOCABULARY LIST B
Directions: Study the vocabulary. Write the missing words in the blank spaces.

1. grit size   the size of the abrasive grains in the grinding wheel. *Example:* A
_____ _____ of 50 is more coarse than 100 or 500.

2. brittle   easily broken or shattered because it is hard and will not bend.
Example: Glass is an example of a _____ material.

3. structure   how something is built; the arrangement of the parts within the whole.
Example: The _____ of a grinding wheel tells how closely
the abrasive grains are: *open, medium,* or *dense*.

4. vitrified   changed into a glass-like substance by a process of high-heat fusing.
Example: This wheel uses a _____ bond.

5. resinoid   like resin, which is a liquid coming from plants or trees and which can be
hardened. *Example:* Bui bought a wheel with a _____ bond.

6. rim   the edge, often of something circular. *Example:*
The _____ of the wheel is the periphery.

**wheel shape:
flared cup**

7. to flare   to curve outward like the rim of a bell. *Example:*
It is possible _____ the rim of the wheel
when a cup-shaped grinding wheel is being made.

14-08: READING
Directions: Read the following paragraphs.

MORE ABOUT SELECTING A GRINDING WHEEL

2. Grit size: This is the size of the abrasive grains in the grinding wheel. Coarse grains have low numbers, such as 4, and fine grains have high numbers, like 500. Commonly used wheel grits are from 46 to 100. Coarse grits are for soft material and fine grits for hard, brittle material.

3. Hardness: In addition to the hardness of the abrasive, the word *hardness* refers to the strength of the bonding material holding the grains together. If the bond is stronger, the wheel will be harder. The hardest bond possible is not always wanted; sometimes it is better to have a softer bond, so that old grains at the surface can be more easily pulled away, to expose new ones below.

4. Structure: The spacing of the abrasive grains in the wheel are given numbers from 1 (for *dense*) up to 15 (*open*), with *medium* in between. Structure supplies clearance for the cutting action of the grains. This keeps the wheel surface from becoming loaded with small particles of metal.

5. Kinds of Bond: The textbook describes 5 kinds of bond and their abbreviations used in identifying different wheels. (1) **vitrified (V)** bonds are the most common and can be used for precision grinding; (2) **resinoid (B)** bonds are used for rough grinding and the removal of large amounts of stock; (3) **rubber (R)** bonds are used with centerless grinders; (4) **shellac (E)** bonds are used in a limited way for some tasks like finishing auto camshafts; (5) **metal (M)** can be used as the bond for diamonds, Borazon, and other super abrasives; metal bonded wheels can be used for grinding hard non-metallic stone and ceramics.

14-09: EXERCISE

Directions: For each statement, circle T for "true" or F for "false."

T F 1. A grit size of 65 will be finer than a grit size of 150.

T F 2. Coarse grits are used on hard, brittle material, but not on soft material.

T F 3. For the best grinding work, it is always best to have the hardest possible bond.

T F 4. The spacing of the abrasive grains can help with clearance for the grinding wheel.

T F 5. Sometimes the cutting face of a wheel becomes "loaded" with tiny metal chips.

T F 6. Bonds made of shellac are the most commonly used for making grinding wheels.

T F 7. Diamond, Borazon and other superabrasives are usually held by metallic bond.

T F 8. Resinoid bonds are used for rough grinding and the removal of large amounts of stock.

14-10: CLASSROOM LECTURE   *BE SAFE WHEN YOU GRIND*

Directions: Listen to Al Lopez as he talks to the students about grinding machine safety. Fill in the blanks with what you hear on the tape. Practice your pronunciation. Listen again and practice taking your own notes on another sheet of paper.

Al:  During our months together we've _____ about _____ frequently, and now we need to _____ it again, because we're starting to learn about the _____ machines. Some of the _____ you learned for _____ safety apply here. Be sure to wear safety _____ with good wide side _____--there's a lot of very fine _____ coming off the wheel.

Next, be sure to learn the _____ and purpose of all the _____ on any machine you're using. We have grinders from different _____ and the _____ of the controls is a little _____ on each machine. Know where the motor ON/OFF _____ are, and, by the way, many grinding machines have more than one _____ that will be running at the same time. They each have their own controls.

Also remember to _____ properly for work. Avoid anything that might get caught in the _____ parts--long sleeves, long _____, ribbons, jackets, _____ , watches, _____ , necklaces.

The machines use a lot of _____, and it can be quite dangerous if it _____ out on the floor--you've got a very slippery deck then. So keep the floors mopped up.

Le: Let me see if I understand what you've said. There are some safety rules we need to follow on these grinding machines. Like always, we need to wear safety glasses and we avoid wearing anything that can get caught in the moving parts. We need to learn the individual machines, especially where the controls and ON/OFF switches are. And we need to wipe up any spills.

14-10: CLASSROOM LECTURE

Al: That's a good summary, Le. Now I want to show you _____ to do a ring test before you mount a grinding wheel on the machine. You want to make sure you're not putting a cracked _____ on the machine, because it will break _____ and the _____ go flying! Here's how you do it: First, look for _____ in the wheel--sometimes they're visible. To make sure, hold the wheel with the _____ of your finger through the center bore; then _____ the wheel lightly with a screwdriver handle or a non-metallic mallet. If the sound is dull, the wheel is _____ and shouldn't be used. If you hear a clear ringing sound, the wheel is _____ cracked and you can use it. Here watch me do it.

Does the wheel ring?

mallet

A Ring Test

14-11:   READING AND DRAWING
Directions: *Read the following paragraphs that summarize the information on page 649 in the textbook. Then draw a picture of the cross-section of each grinding wheel; then label the important dimensions of each.*

6. Size and Shape of the Wheel: Numbers 1 to 28 are assigned to indicate different sizes and shapes of grinding wheels. The textbook illustrates five of the most common types:

Type 1, the straight wheel: grinding face is the periphery; dimensions vary according to outside diameter of the wheel (D), hole diameter (H), and wheel thickness (T).

Type 2, the cylinder wheel: grinding face is the rim of the wheel; important dimensions are the outside diameter of the wheel (D), the wall thickness (W), and wheel thickness (T).

Type 6, the straight cup wheel: grinding face is the rim of the wheel; dimensions are wheel diameter (D), hole diameter (H), wall thickness (W), and wheel thickness (T).

178

14-11: READING AND DRAWING

Type 11, the flaring cup wheel: grinding face is the rim of a tapered cup; the flaring cup has a larger outside diameter (D), a smaller outside diameter (J), a bottom inside diameter (K), a wall thickness at the rim (W), and a rim-to-bottom wall thickness.

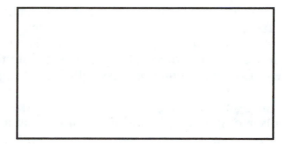

Type 12, the dish wheel: grinding faces are the rim of the cup and the narrow, straight periphery of the dish; in additition to the dimensions given for Type 11, this type has a rim inside diameter (M) and a periphery thickness (U).

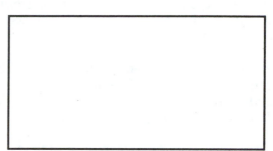

14-12: VOCABULARY LIST C
Directions: Study the vocabulary. Write the missing words in the blank spaces.

1. glazed — having a hard, glass-like finish, said of grinding wheels whose cutting surface is loaded with chips and is smooth. *Example:* The aluminum oxide wheel didn't work well because its cutting surface was _____.

2. loaded — said of a grinding wheel when its cutting surface is filled with tiny chips of metal. *Example:* The silicon carbide wheel was _____ and would not cut . (SDCCD)

3. cluster — a number of things of the same kind gathered together; it is said here of diamonds held in a metal bond. *Example:* A _____ can be used to break the abrasive grains on the surface of the grinding wheel.

4. flange — a rim which sticks out from a wheel to hold it in place, and guide it. *Example:* Do not tighten the nut on the the wheel too much; you can damage the _____ s and even crack the wheel.

inner flange   outer flange

14-13: READING
Directions: Read the following paragraphs.

TRUING, DRESSING AND BALANCING A GRINDING WHEEL

To work efficiently, a grinding wheel must occasionally be reshaped so that its periphery is concentric at all points with the spindle of the machine; this process is called **truing**. It is done when the wheel is new, and it is done when the wheel's periphery loses its concentricity through use and the development of high and low spots on that outside edge.

179

14-13:   READING   (continued)

With use, especially if the machinist makes a poor choice about the type of wheel to use, the cutting faces can become filled with small metal chips, the wheel can become glazed, and fresh grains may not be coming to the surface.   Then the *loaded* wheel must have the metal chips removed from the cutting surfaces of the wheel in a process called **dressing**.

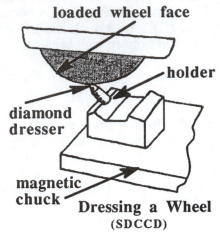

Both truing and dressing use a tool called a **dresser**, which is usually a diamond, either **single-point**, or in a bar called a **cluster**.   The dresser is brought into contact with the rotating wheel to cut away grains and bond from the cutting surface; the single-point dresser will traverse across the width of the cutting face.

The process of dressing renews a loaded wheel by exposing new grains.   The process of truing the wheel removes high spots and continues until the dresser is removing material from all around the cutting face.

Dressing a Wheel
(SDCCD)

A grinding wheel may be concentric with the spindle and have a well-prepared cutting surface, but it may still leave chatter marks because the weight of the wheel is not evenly distributed.   In this case the wheel will need **balancing**.   That is done by adding weights to the flanges opposite the spot where the extra weight occurs.   This is necessary on wheels with diameters over 14 in.

14-14:   "SHOP TALK"   EXPLAINING A PROCEDURE

Directions: Listen to the tape and fill in the missing words.   To explain a procedure, use words like "first, second, third," and "next, then, finally."

Teacher: Tell us the _____ for dressing and truing a grinding wheel.

Student: To dress a grinding wheel, _____ mount it on the machine_____.

               _____, position a diamond dresser in its holder so that it touches the wheel at the point of outfeed, that is beyond the _____ where the wheel usually _____ the work.

               _____, turn on the wheel and begin to _____ it down into the dresser point.

               _____, give some traverse so the _____will contact the full width of the wheel.   This can be done with the table feed handwheel.

Teacher: Yes, that's good.   Tell us more about _____ the wheel.

Student: Truing goes beyond dressing, because you may have to take _____ off the periphery to make the wheel truly _____ with the spindle.   In truing you're not just sharpening the face of the wheel; you're also _____ it to make it round.

180

14-15: VOCABULARY LIST D

Directions: *Study the vocabulary.   Write the missing words in the blank spaces.*

1. pedestal

a column which supports something above it. *Example*: The ABC Machining Co. won the inter-business softball championship. They put their trophy on a _____.

trophy

(SDCCD)

pedestal

2. trough

a long, narrow, open container for liquid or other loose material. *Example*: The coolant ran off the wheel and the workpiece into the coolant _____.

3. vacuum

the suction provided to remove chips and coolant from the grinding interface. Air rushes into an empty place and carries unwanted material with it. *Example*: The new grinding machine has _____ hoses which carry away unwanted swarf and coolant.

4. fitting

a small part used to join or adapt other parts, as in a system of pipes. *Example*: This pedestal grinder has a _____ to which I can attach a vacuum hose for the removal of coolant and swarf.

5. to ring

to tap a grinding wheel and listen for the clear sound of an uncracked wheel. *Example*: It is possible _____ a grinding wheel to see if it is cracked or not.

6. to reciprocate

to move back and forth, first in one direction and then in the exact opposite. *Example*: The table of a surface grinder must be able _____ with the work on it, as the wheel is lowered into the work.

7. magnetic

the ability of a piece of iron or steel to attract other pieces of iron or steel; the force of mag-netism can be strong enough to hold heavy pieces in place; magnets can be permanent or they can be electromagnetic, activated by electrical power. *Example*: A common work-holding device used with reciprocating tables is the _____ chuck.

table T-slots

T-slot clamp

magnetic chuck

activating handle

8. to deflect

to make something go to one side or to change course. *Example*: The deflector on the grinder is able _____ any flying pieces and prevent injury in the shop.

9. focus

the center of attention; the main effort or idea. *Example*: The _____ of this lesson is to learn all we can about grinding machines.

10. foundation

the base on which something rests. *Example*: The cast iron bed makes a good _____ for the table and the other parts of the grinder.

Directions: Read the following paragraphs.

KINDS OF GRINDING MACHINES

1. The Pedestal Grinder:
Grinding machines used in the machine tool field vary in size from small **bench grinders** to huge industrial grinders (See Figure N-7 in the textbook). A smaller machine often found in the machine tool classroom is the **pedestal grinder**. It is useful for sharpening drills, toolbits, center punches, and reshaping other small tools like screwdrivers. The machinist holds the tool in the hand and feeds it into the rotating wheel. Students need to develop skill in these hand-held practices.

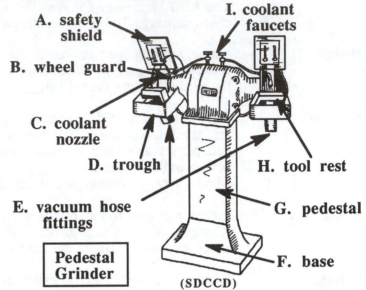

A. safety shield
I. coolant faucets
B. wheel guard
C. coolant nozzle
D. trough
H. tool rest
E. vacuum hose fittings
G. pedestal
F. base

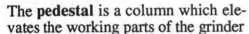

Pedestal Grinder

(SDCCD)

The **pedestal** is a column which elevates the working parts of the grinder above the **base**. In the picture are two **grinding wheels** which are turned by the **motor** in the center. When the machine is running, the machinist can hold whatever needs sharpening on the **tool rest**; it is there to support and steady the grip in this hand-held operation. **Coolant faucets** can be turned on to spray coolant on the work through **coolant nozzles**. Used coolant is collected in a **trough**. A clear plastic **safety shield** protects the operator, but caution is still needed, and safety glasses should be worn; the **wheel guards** cover most of the wheel as another safety feature. Vacuum hoses can be attached to the **vacuum hose fittings** in order to suction away the swarf and cutting fluid. The bench grinder is like the pedestal grinder, but it is mounted on a workbench instead of a pedestal.

Whenever a grinding wheel needs to be replaced on this or other machines, it is important to get the correct size and to do a **ring test** on the new wheel. This is done by holding the wheel with a finger, tapping the wheel with a piece of wood, and listening for a clear, ringing sound. The ringing sound tells the machinist that the wheel is not cracked and is safe to use. A cracked wheel will give out a dull sound when it is tapped. Further details about hand grinding are given in Section B, Unit 9 of the textbook, pages 78 to 81.

14-17: EXERCISE
Directions: For each statement, circle T for "true" or F for "false."

T F 1. The pedestal grinder is used mainly in industry for finishing large numbers of parts.

T F 2. The main use for a pedestal grinder in the classroom shop is sharpening tools.

T F 3. If the pedestal grinder is equipped with a safety shield, you won't need safety glasses.

T F 4. Vacuum hoses are attached to the grinder to remove swarf and cutting fluid.

T F 5. The cutting fluid and the coolant are the same thing.

T F 6. It is important to handle any grinding wheel with care, because it can become cracked.

14-18: READING
Directions: Read the following paragraph.

A MUCH-USED GRINDING MACHINE

2. The Surface Grinder: A larger machine that is used in most classrooms is the surface grinder. There are several types of such grinders, but the most common is the **horizontal spindle reciprocating table surface grinder.** The spindle of this machine lies in a horizontal plane; the table reciprocates beneath the grinder; and the main purpose of the machine is to grind flat surfaces. Learn the parts.

MAIN PARTS:

A. downfeed handwheel
B. grinding wheel
C. coolant hose
D. magnetic chuck
E. coolant trough
F. table reverse trip
    dogs
G. automatic controls
H. cross feed
    handwheel
I. table feed handwheel
J. saddle
K. reciprocating table
L. deflector
M. wheelhead
N. lamp
O. downfeed ways

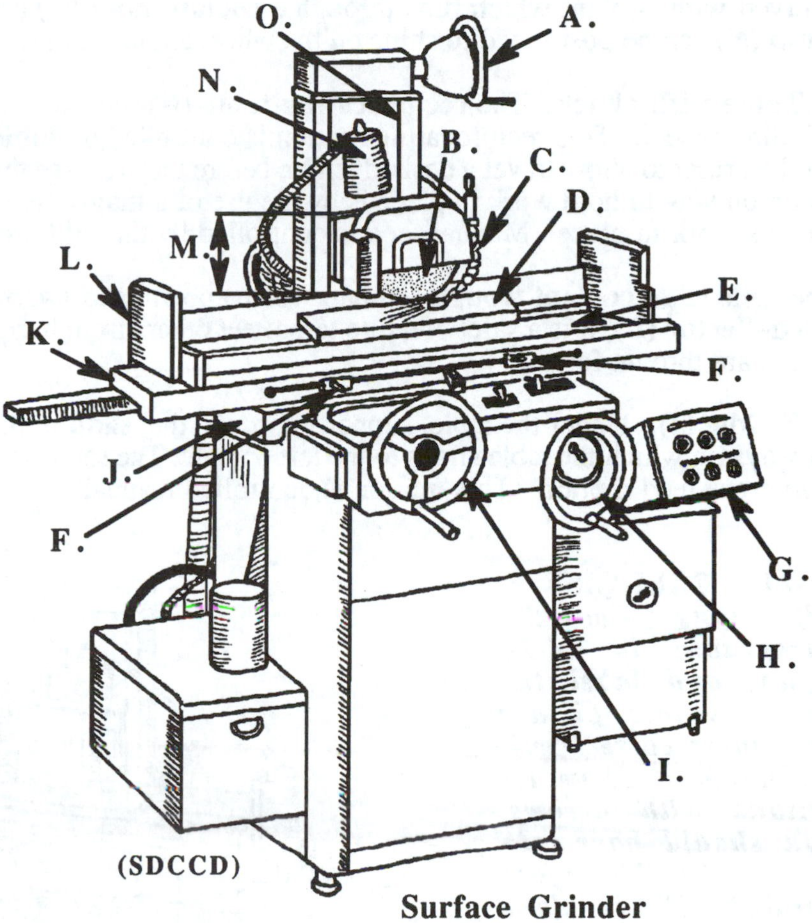

14-19:
NOMENCLATURE
Directions:
Look at the picture. Say the names. Listen to the tape, and practice your pronunciation.

Listen several times; then listen and try writing the names in the spaces.

(SDCCD)

Surface Grinder

A. _____

B. _____

C. _____

D. _____

E. _____

F. _____

G. _____

H. _____

I. _____

J. _____

K. _____

L. _____

M. _____

N. _____

O. _____

14-20: READING

Directions: Read the following paragraphs. The letters after the part names are the labeling letters from the previous page.

SURFACE GRINDER PARTS

a. Wheelhead (M): This part holds the driving **motor** and the **spindle** on which the **grinding wheel (B)** is mounted. The wheelhead slides up and down a set of vertical **down-feed ways (O)**, so the grinding wheel can be lowered into the work. Most machines will have both an **automatic control (G)** and a **downfeed handwheel (A)**. The wheel/work interface is sprayed with coolant which runs through a **coolant hose (C)** and out a **coolant nozzle**. A **lamp (N)** can be positioned to shine on the wheel and the work.

b. Table with chuck: The **reciprocating table (K)** moves to the right and the left under the grinding wheel. This reciprocating motion is controlled by **table reverse trip dogs (F)** which can be preset to allow travel a desired length before they reverse the direction of the stroke. A common way to hold work in place is by the use of a **magnetic chuck (D)** which will hold ferrous work in place. Manual feed is controlled by the **table feed handwheel (I)**.

The table has a **coolant trough (E)** into which coolant and swarf runs. At the end of the trough is a **deflector (L)**; it is a safety feature to protect others from harm if a grinding wheel breaks and pieces are thrown forward.

c. Saddle (J): Under the table, supporting it, is the **saddle**; the top of the saddle has horizontal **ways** on which the table slides to the left or right. The saddle itself slides in and out on ways to give cross feed motion. This motion is controlled manually by the **cross feed handwheel (H)**.

14-21: EXERCISE

Directions: Read the paragraphs in 14-20 again, and label the parts in this picture with their correct written names. Show locations with arrows. You should have 15 parts. The "downfeed handwheel" is done as a sample.

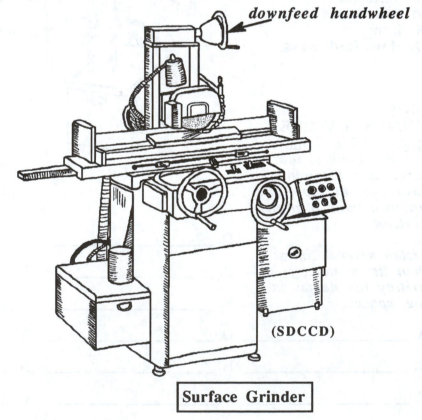

downfeed handwheel

(SDCCD)

Surface Grinder

184

14-22: EXERCISE
Directions: _Choose the best answer by circling the letter._

1. The surface grinder part which slides up and down on vertical ways is the:
      a. reciprocating table   b. saddle   c. wheelhead   d. deflector

2. A frequently used work-holding device on the surface grinder is a:
      a. magnetic chuck   b. 3-jaw chuck   c. 4-jaw chuck   d. drive plate and dog

3. The manual control that feeds the grinding wheel into the work is the:
      a. table feed handwheel   c. cross feed handwheel
      b. downfeed handwheel   d. deflector

4. Which one of these features of the surface grinder is primarily for safety?
      a. the reciprocating table   c. the trip dogs
      b. the coolant hose   d. the deflector

5. What causes the table to reciprocate?
      a. the saddle   b. the wheelhead motor   c. the trip dogs   d. the deflector

14-23: MACHINE NOMENCLATURE   _AUDIO QUIZ_

Directions: _Test yourself on the names of the parts of the surface grinder. Look at the picture. Listen to the names on the tape. Write the letter of what you hear next to the number._

1. ___
2. ___
3. ___
4. ___
5. ___
6. ___
7. ___
8. ___
9. ___
10. ___
11. ___
12. ___
13. ___
14. ___
15. ___

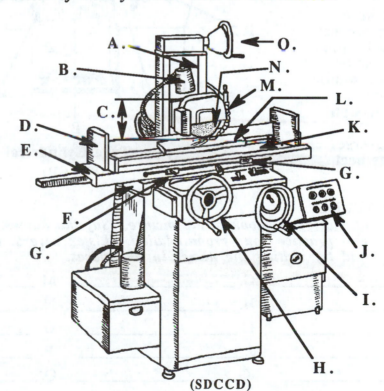

(SDCCD)

185

Directions: *Read the following paragraph.*

THE CYLINDRICAL GRINDING MACHINE

3. The Cylindrical Grinder: The textbook on pages 633 to 639 explains and illustrates a number of grinders that are used to produce round work of various kinds; included are the center-type grinder, the roll-type grinder, the centerless grinder, the internal cylindrical grinder, and the tool and cutter grinder. This reading will focus on the **center-type cylindrical grinder**. Study its main parts, so you will be able to ask and answer questions about this useful machine.

MAIN PARTS:

A. table reverse lever
B. automatic controls
C. trip dogs
D. bed
E. cross feed handwheel
F. swivel table
G. footstock
H. wheelhead
I. internal grinding head
J. internal grinding wheel
K. coolant hose
L. grinding wheel
M. headstock spindle with center
N. headstock motor
O. headstock
P. coolant trough
Q. slide
R. table traverse handwheel

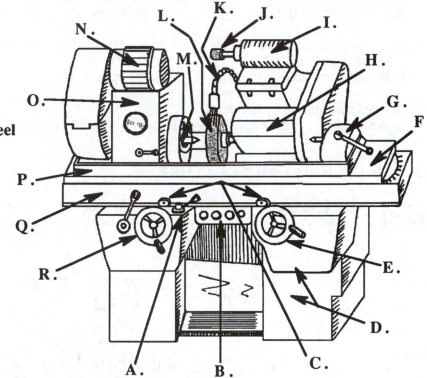

Center-Type Cylindrical Grinder
(SDCCD)

14-25:
NOMENCLATURE

Directions: *Look at the picture. Say the names. Listen to the tape, and practice your pronunciation. Listen several times; then listen and try writing the names in the spaces.*

A. _____   G. _____   M. _____

B. _____   H. _____   N. _____

C. _____   I. _____   O. _____

D. _____   J. _____   P. _____

E. _____   K. _____   Q. _____

F. _____   L. _____   R. _____

Directions: *Read the following paragraphs. The letters after the part names are the labeling letters from the previous page.*

CYLINDRICAL GRINDER PARTS

A machine is called *universal* when it has added features which allow it to work in more positions than what a basic machine can do. A **universal center-type cylindrical grinder** is called *universal*, because the wheelhead of the machine is able to swivel in the horizontal plane; on a plain machine the wheelhead can not swivel, but traverses straight in and out on ways that are at 90° to the table. Study the parts of this grinder.

a. Bed (D): It is the strong cast iron foundation for the other parts of the machine. The bed supports the slide and the wheelhead.

b. Slide (Q): the slide runs over ways cut into the bed. Its traverse motion carries the work past the wheel; the traverse can be reciprocating when the **trip dogs (C)** contact the **table reverse lever (A)** in their right and left movements. The traverse of the slide is controlled by the **table traverse handwheel (R)** or by the **automatic controls (B)**.

c. Swivel table (F): It rests on the slide and can be swiveled in the horizontal plane, up to 30° in either direction, to make possible the machining of tapers of different angles. The swivel table has T-slots for attaching the headstock and the footstock.

d. Headstock (O): This part attaches to the swivel table and supports one end of the work. The headstock has its own **motor (N)** which turns the **headstock spindle (M)**. This spindle will hold a center for work between centers; on it may also be mounted chucks with 3 or 4 jaws, or a face plate, or a round magnetic chuck. The purpose of the headstock is not only to support the work but to rotate it, so the grinding wheel will grind the entire circumference of the work. When a center is used in the spindle, a driving plate and dog are used to grip and rotate the workpiece.

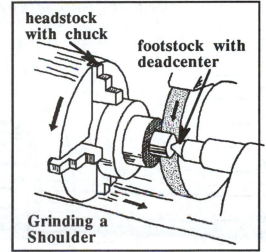

headstock with chuck

footstock with deadcenter

Grinding a Shoulder

e. Footstock (G): It is also mounted on the swivel table. It has a spring-loaded dead center which holds the end of the work opposite the headstock; this center does not rotate the work. The footstock can be positioned anywhere along the T-slots of the swivel table for holding different sized pieces.

f. Wheelhead (H): This part holds its own driving **motor** and the **spindle** on which the **grinding wheel (L)** is mounted. The wheelhead slides in and out on horizontal ways and swivels on a turret base, so the grinding wheel can be fed into the work. Most machines will have both **automatic controls (B)** and a manual feed, the **cross feed handwheel (E)**. The wheel/work interface is sprayed with coolant which runs through a **coolant hose (K)** and is collected in a **coolant trough (P)**.

g. Internal grinding head (I): The machine is also capable of grinding inside diameters by the use of an internal grinding head, which can be swung back out of the way when not in use. The grinding head has its own motor and can take a variety of **internal grinding wheels (J)**.

14-27: EXERCISE
Directions: *Choose the best answer by circling the letter.*

1. The foundation, on which all the other parts of the cylindrical grinder rest, is called the:
    a. footstock   b . headstock   c. wheelhead   d. bed

2. What can be used to hold the work on a cylindrical grinder?
    a. magnetic chuck   b. 3-jaw chuck   c. drive plate and dog   d. all of these

3. Which one of these parts contains a motor?
    a. headstock   c. wheelhead
    b. internal grinding head   d. all of these

4. Which part on the cylindrical grinder makes it easy to grind a tapered shaft?
    a. the swivel table   c. the trip dogs
    b. the internal grinding head   d. all of these

14-28:   MACHINE NOMENCLATURE   *AUDIO QUIZ*

Directions: *Test yourself on the names of the parts of the cylindrical grinder. Look at the picture. Listen to the names on the tape. zWrite the letter of what you hear next to the number.*

1. ____
2. ____
3. ____
4. ____
5. ____
6. ____
7. ____
8. ____
9. ____
10. ____
11. ____
12. ____
13. ____
14. ____
15. ____
16. ____
17. ____
18. ____

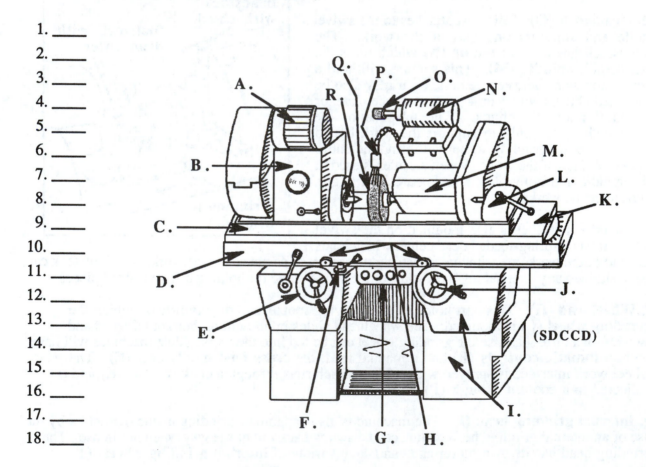

(SDCCD)

188

Unit 15: *CNC MACHINING*

ASSIGNMENT: **Read, study and complete pages 189 to 198 of this book. Then read pages 707 to 753 in the textbook,** *Machine Tool Practices,* *SECTION O, Advanced Machining Processes.*

OBJECTIVES for this unit:
You should be able to:
1. Pronounce correctly and answer questions related to some CNC machining vocabulary.
2. Identify the names and functions of some common parts of a computer.
3. Correctly position points in a rectangular coordinate system.

Important note to the student: In the previous chapters you have studied many basic areas of machining and the language that you will need to succeed on the job. Now take a quick look at some of the key ideas and language used in CNC machining. In most schools you will find that CNC is an advanced course which you can take after you master the basics.

15-01: VOCABULARY LIST A
Directions: *Study the vocabulary.* *Write the missing words in the blank spaces.*

1. recent   done just before the present time, new, modern. *Example*: The toolroom is a _____ addition to the machine shop.

2. advance   an improvement, making something better so it can be done more efficiently, or with less cost. *Example*:   Computer control of many machining operations is an _____ over the machinist's doing everything.

3. sequence   a continuous series of actions in which step one comes before step two, two before three, etc. *Example*: A _____ of steps can be prorammed into the computer; they will be carried out one after another.

4. to regulate   to control, to direct, to adjust something so it works accurately.
Example: Before CNC, the machinist had _____ all the controls on any machine he/she used.

5. execution   doing, acting on a plan. *Example*: Juan is responsible for the planning and _____ of all the procedures on his lathe.

6. program   a set of directions for doing something, written in a language that a computer-controlled machine can understand and execute. *Example*: This _____ tells the drill press to drill five evenly-spaced holes.

```
600  START  INS  02
601  TO=        0.2500
602  FR  XY    =20.0
603  FR        Z=02.0
```
A CNC program begins.

7. toolpath   a line along which a tool moves during a machining operation. *Example*: Sonja programmed the end mill to move along a _____ of 5.5000 inches before it started cutting out the pocket.

189

8. durability   the quality of lasting a long time, even with hard work and frequent use. *Example*: Because of new alloys, some tools have greater _____ than they did a few years ago.

9. traditional   being done by methods that have been passed on to us from earlier times. *Example*: Cutting metal with a laser may replace some _____ methods of cutting metal with machine tools.

10. ultra-sonic   related to vibrations that are of higher number than those we can hear under usual conditions. *Example*: Zoe learned that machinists can shape metal using _____ vibrations.

15-02: READING
Directions: ***Read the following paragraphs.***

RECENT ADVANCES IN THE FIELD OF MACHINING

The basic goal of machining has not changed: to process materials, like metal, in order to make a variety of useful products. However, in recent years the machining trade has made some important changes.

a. Control of machines: Previously, the machinist planned the steps of an operation and the sequence of machines and tools to be used for a particular job, as described in a blueprint. The machinist also regulated speeds, feeds, and depths of cut, with manual and automatic controls; Now, more and more, the planning and execution of the operations have been given over to computer programs and to computer programmers in a process called **computerized numerical control (CNC)**.

b. Efficiency of tools: Advances are being made in the tools used to cut materials: With toolpaths precisely planned and tools used in correct sequence and under best conditions, work can be shaped at higher speeds and feeds than has been possible with the cautious approach of manual work.

New materials and new tool geometries are adding to the efficiency of tools.

There are continual improvements in the quality of tools, both in their tool geometry and the durability of the materials used to make them.

There are also a range of non-traditional tools being used which are not based on traditional chip-removing technology; laser cutting tools, steel-cutting water jets, ultra-sonic machining, and electric spark (EDM) machining are some of the methods used. (See pages 748-753 in the textbook.)

c. Greater productivity: The CNC machine reduces or eliminates the time a machinist needs to think about the next step, check tooling, measure the work, and many other steps of manual machining; the CNC controlled machine works continually, following over and over the same directions from the computer without need to rest or eat lunch.

15-03: EXERCISE
Directions: _Use complete sentences to write short answers to this question._

Question: Describe three recent advances in the field of machining.

1. _____

_____

2. _____

_____

3. _____

_____

15-04: VOCABULARY LIST B
Directions: _Study the vocabulary. Write the missing words in the blank spaces._

1. complex — made up of two or more interrelated parts. _Example_: A computer is a _____ machine.

2. keyboard — a plastic set of keys on a single board which allows information and instructions to be typed into a computer. _Example_: Abdullah used the _____ to change the program for running that part.

Computer Keyboard

3. magazine — a place for storing things until they are needed. _Example_: That machining center has a _____ with many tools loaded into it.

4. drum — a cylinder-shaped storage place for tools, which can be rotated into place when they are needed. _Example_: Figure O-15 in the textbook shows a _____ with tools loaded around its periphery.

5. adaptive — able to change to take care of new or developing situations. _Example_: Some CNC machines have _____ controls which will automatically change the feed rate as conditions change.

6. to monitor — to check on the performance of a machine and make adjustments if needed. _Example_: A quality control inspector comes to my work station twice a day _____ my work.

7. code — a set of symbols or words used to control the work of a computer. _Example_: Some CNC machines get instructions on what to do in a machining operation from a G _____ and from an M _____.

8. to network — to join two or more computers so they can communicate with each other and share programs. _Example_: At the factory the engineers were able _____ the CNC machines with a master computer.

Directions: *Read the following paragraphs.*

A FEW FEATURES OF CNC MACHINES

a. Variety of CNC machines: There are a variety of machines that are contolled by computer programs: A machine might be a simple, table-top, computer-controlled drilling machine, or it might be a huge **machining center** with a variety of tools, a large capacity, and the ability to follow many complex steps. (See the pictures on pages 707-708 and 718-720 of the textbook.) In between these sizes are the machines we have studied (drills, lathes, milling machines, grinders) with all the features of those machines, but controlled by computer programming.

b. Machine control unit (MCU): The program which guides the operations is usually written away from the actual machine that does the work; however, the CNC machine has a machine control unit into which the program is loaded and which includes a computer keyboard for making small changes or corrections in the program.

c. Automatic tool changers (ATC): The program includes instruc-tions on what tools to use for the different part-making operations. Therefore, on multi-tool CNC machines, the tools must be loaded into the machine so they can be easily swung into place when they are needed. Figure O-12 shows a **tool magazine**; Figure O-15 shows a **tool drum**; and Figure O-18 shows tools ready for turning, mounted on a **tool turret**.

tool turret

d. Adaptive control (AC): Some CNC machines have controls that are constantly monitoring conditions during the machining operations and are making adjustments to get the most efficient use from the machine. Things that the adaptive control can monitor and adjust include feed rate and tool failure.

15-06: EXERCISE
Directions: *For each statement, circle T for "true" or F for "false."*

T F 1. The use of CNC is increasing in the machine tool industry.

T F 2. The operator must refer all adjustments to the program back to the programmer.

T F 3. An adaptive control on a CNC machine can adjust the feed rate without the operator.

T F 4. Each time a new tool is needed, the CNC operator must stop the machine to change.

T F 5. Only tools with vertical spindles can be controlled by a computer.

15-07: READING
Directions: *Read the following descriptions of a basic computer system.*

CNC machines are controlled by **computers**. Before going further, study (on this page and the next) some of the basic nomenclature that is used to talk about computer systems.

The heart of any computer is the **central processing unit** (CPU); it is often contained in its own housing and consists mainly of a **computer chip** which works with information coming in and going out of the computer. The computer information is displayed on the **computer screen**.

The computer screen is mounted on the front of the **monitor**, which is much like a TV set.

For putting information into the computer, a **keyboard** allows the user to type and see what is written on the screen; some computers have a **mouse** which is used to point to information and to features of the computer's programs; the mouse is rolled on a **mouse pad** to perform its work.

The computer has the ability to store programs, as well as information which it produces or which it receives from other sources. This ability is the computer **memory**; the memory the computer uses to run the programs is called **RAM (random access memory)**; and the memory used for storage is called **ROM (read only memory)**. Much of the memory is located inside the CPU in the **internal hard drive**, the storage space for memory. Programs and infomation can also be saved on floppy disks or compact discs. Most CPUs have slots **(external drives)** for inserting floppies and CDs. Memory is measured by these units: **byte, kilobyte** (1,000 bytes), **megabyte** (1 million bytes), and **gigabyte** (1 billion bytes).

The various parts of a computer system are connected with computer **cables**.

A computer program is a set of instructions that tells the computer what to do. A computer's collection of programs and information-storing devices like floppy disks are called **software**. The CPU, the monitor, and the other large pieces of the system are called computer **hardware**.

15-08: NOMENCLATURE OF THE COMPUTER

Directions: *Study the pictures as you listen to the nomenclature of a small computer system. Practice pronouncing the words.*

A. keyboard
B. mouse pad
C. mouse
D. cable
E. screen
F. monitor
G. external drives
H. internal hard drive
I. CPU (central
      processing unit)

Main Parts of a
Computer System

Write the names as you listen.

A. _____

B. _____

C. _____

D. _____   G. _____

E. _____   H. _____

F. _____   I. _____

15-09: READING
Directions: Read the following paragraphs.

KEY PERSONS IN A CNC ENVIRONMENT

In larger shops and factories, the jobs are divided among three kinds of workers:

a. Computer programmer: This person writes the directions which tell the computer what tools to use and in what order, what paths to follow, when to start and when to end, and all the other steps needed to make the part. The programmer will use **codes** that the computer can under-- stand to control what the machine does.

The programmer must be familiar with the computerized machinery of many different companies, each of which may have some different code words, though many will be the same from one machine to the next. As the person who plans the operation of the machine, a programmer should keep learning more and more about new materials, new tools, and new methods as they come into the field.

Writing a CNC program

b. Setup person: This worker loads the program into the MCU, either directly or from a master computer which is networked with several machines and which holds many programs. The setup person also makes sure that the machine is loaded with the tools needed for the operations and that the tools are in good condition. The setup person runs a first part to check that the program is running as planned. Needed adjustments to the program can be made at the MCU.

c. Machine operator: The operator loads pieces of rough stock into the machine work-holding devices, removes finished parts, and watches for things like tool wear or breakdown.

15-10: EXERCISE
Directions: Read the following descriptions of a person and write down one of the following workers: PROGRAMMER, SETUP PERSON, CNC OPERATOR.

1. I load rough stock into the CNC machine. Who am I? _____

2. I write codes which the CNC machine uses to guide its movements.

    Who am I? _____

3. As the work is being done, I must watch for tool wear and tool breakdown.

    Who am I? _____

4. I make sure that all the tools needed for the job are loaded into the CNC machine.

    Who am I? _____

5. I run the CNC machine to make the first part; then I make adjustments to the program, if they

    are needed. Who am I? _____

6. I am responsible, more than others, for keeping up with information about recent advances in

    tool technology and computerized equipment. Who am I? _____

194

15-11: VOCABULARY LIST C
Directions: Study the vocabulary. Write the missing words in the blank spaces.

1. pair — two things or two parts which are used together. *Example:* I can locate a point on the map by giving you a _____ of numbers.

2. coordinate — any one of two or more numbers used to tell the position of a point, line, curve, or plane. *Example:* Give me the first _____ for locating the center of that hole.

3. Cartesian coordinates — a pair of numbers that locate a point by its distances from two intersecting perpendicular lines in the same plane, the X axis and the Y axis. *Example:* Sometimes a CNC machine uses _____ _____ to locate a point for a drill to make a hole.

4. quadrant — a quarter section of a circle. *Example:* Four _____s are formed when two perpendicular lines intersect in a plane.

5. to plot — to mark the location of a point by the use of coordinates. *Example:* I have _____ the locations of the centers for this five-hole bolt circle.

15-12: READING
Directions: Read the following paragraphs.

HOW A CNC MACHINE USES NUMBERS

CNC machines routinely produce parts with high dimensional accuracy and close tolerances. The program codes, like a **G code** and an **M code**, tell the CNC machine what operations to perform, what tools to use, and what dimensions to give to the features of the part. Each line of code has a number, with numbers for each tool used, numbers in inches or millimeters for dimensions and tolerances, and numbers for all other parts of the operation.

One way of telling a machine what path a particular tool should travel is to use a system of **rectangular coordinates** (sometimes called **Cartesian coordinates**). A rectangular coordinate system is made up of two axes which are perpendicular to each other and which lie in the same plane; one axis is called the **X-axis** and the other is the **Y-axis**; the two axes intersect at a point called the **origin**. Each arm of both axes is divided into numbered units: units on the X axis which lie to the right of the Y axis are **positive X values**; units on the X axis which lie to the left of the Y axis are **negative X values**; units on the Y axis above the X axis are **positive Y values**, those below are the **negative Y values**. All units are measured from the origin out along the four arms.

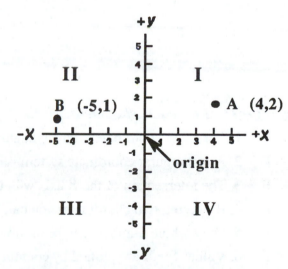

A rectangular coordinate system with its X axis and Y axis intersecting at the origin and having four quadrants

195

The two axes divide the plane into **four quadrants** which are numbered in a counterclockwise direction starting at the upper right with the numbers **I, II, III, IV**.

Such a coordinate system is useful in machining because it can be used to locate points in the plane by giving a **pair of numbers** (x,y) in which the x value is given first, and the y value is given second. The origin is then designated as (0,0) and all points are measured from that point. In the picture, the point A has coordinates x = 4 and y = 2; the point B, which lies in Quadrant II, has coordinates x = -5 and y = 1. (All positive values are written without a plus sign, but negative values have a minus sign.

15-13: EXERCISE

Directions: Listen to the tape. Write down the coordinates you hear for each letter. Tell what quadrant the point is located in. Then plot the points on the graph. Write the letter for each plotted point.

Name	X	Y	Quad.
A.	3.000	-5.000	IV
B.			
C.			
D.			
E.			
F.			
G.			
H.			
I.			
J.			

15-14: EXERCISE

Directions: For each statement, circle T for "true" or F for "false."

T   F   1. Cartesian coordinates and rectangular coordinates are the same thing.

T   F   2. A rectangular coordinate system is divided into four quadrants.

T   F   3. The intersection of the X axis with the Y axis forms four angles of 180° each.

T   F   4. By giving two coordinates you can locate a point on a plane.

T   F   5. Only whole numbers can be used when plotting points in a coordinate system.

T   F   6. Values for X are plotted by counting units out from the Y axis.

T   F   7. Values for Y are plotted by counting units out from the Y axis.

T   F   8. In a Cartesian coordinate system, the origin is located at the point X0,Y0.

15-15: VOCABULARY LIST D
Directions: Study the vocabulary. Write the missing words in the blank spaces.

1. machining   a computer-controlled machine which has enough tools and positioning.
   center   capabilities to perform several operations. *Example:* Because a _____
                _____ can perform many operations over and over, it is good for
                mass producing parts.

2. "floating"   not fixed in one place; changes locations from one place to another.
                Example: There is one positioning method which has a _____
                origin.

3. incremental the quantity, often a small amount, by which something increases or
                decreases. *Example:* _____ positioning uses a "floating
                zero" or origin which starts at the end of the last toolpath.

4. absolute   unchanged by other changes within a system. *Example:* In _____
                positioning, all points are plotted with reference to one, unchanging origin.

5. to refer   to send something in a certain direction from a given point. *Example:*
                In absolute positioning I am able _____ all points back to one origin.

15-16: READING
Directions: Read the following paragraphs.

A THREE-AXES COORDINATE SYSTEM

The X and Y axes, like a map, are useful for finding particular points on a flat plane. If a third axis is added, unique points in three-dimensional space can be located. This third axis is often called the Z axis. A three axis coordinate system is one of the ways used to locate toolpaths when using CNC machining centers:

a. Types of machining centers: There are two kinds of machining center:

(1) the **vertical spindle machining center (VMC)**; like a vertical milling machine, the VMC has a spindle axis which runs vertically; the machine can be programmed to guide a tool along a toolpath which is laid out with reference to three axes: the **X axis** which runs parallel with the worktable axis; the **Y axis** which is perpendicular to the X axis (in the horizontal plane) and runs parallel with the cross feed axis; and finally the **Z axis** which is perpendicular to the X and Y axes and which lies in the vertical plane.

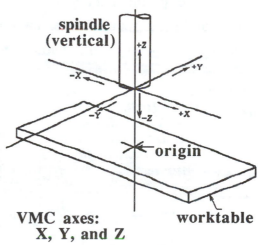

VMC axes: X, Y, and Z

(2) the **horizontal spindle machining center (HMC)**; it is also laid out along three axes: the **X axis** runs with the longitudinal movement of the worktable, as with the VMC, but it has a horizontal spindle axis and the **Z axis** always runs parallel with the spindle axis, so the **Y axis** is used to designate the vertical direction. (For a picture, see page 713 in the textbook)

b. Using the coordinates: With this three-axis system, a CNC program could tell a twist drill which has its point positioned at the origin to move along the positive X axis 3.500 in., along the Y axis -1.000 in. (away from the column), and then to drill down along the Z axis -2.750 in. to make a blind hole in a workpiece which is 5 in. high. Selecting a twist drill of a certain size, the location of the origin, and the movements of the drill along the three axes could all be programmed into a CNC controlled machine.

c. Positioning a tool: There are two ways to position a tool.

(1) in **incremental positioning** a tool starts its path at an origin and moves from there a certain x distance and a certain y distance to the point where it executes its work (for example, drilling a hole); then the point of the drilling becomes a new origin from which the next move will be made to a new location; the tool movement gets its directions each time by referring to its last location. This is sometimes known as a "floating zero" system

(2) in **absolute positioning** a tool's present position is always referred back to its first origin; that will be its only origin; it does not use later positions as origins for its next moves.

Study the examples of these two systems of positioning in Figure O-28 on page 722 in the text and Figure O-30 on page 723.

15-17: MULTIPLE CHOICE   *AUDIO QUIZ*

Directions: *Listen to the tape; circle the letter of the word or phrase which **best** completes the statement or answers the question.*

1. The positioning which uses a floating zero or moving origin is called   A   B   C

2. The spindle axis of a VMC or an HMC always runs parallel with   A   B   C

3. Which one of these axes runs parallel with the worktable axis?   A   B   C

4. What kind of positioning refers its coordinates to an unchanging origin?   A   B   C

5. Which of these is not used to send information to the CPU?   A   B   C

6. The chip for processing information is located in   A   B   C

15-18: NOMENCLATURE

Directions: *Listen to the tape. Look at the picture. Write the letter next to the number.*

1. ____   6. ____

2. ____   7. ____

3. ____   8. ____

4. ____   9. ____

5. ____

ANSWERS TO UNIT QUESTIONS

UNIT 1:   INTRODUCTION TO MACHINE TOOL TECHNOLOGY

pages 2-3, 1-03: 1. *c*, 2. *d*, 3. *c*, 4. *b*, 5. *a*, 6. *a*, 7. *b*
page 3, 1-04: *room, need, core, well, operator, machinist*
page 5, 1-07: 1. *Betty*, 2. *Ramon*
page 6, 1-09: 1. *e*,   2. *d*, 3. *a*, 4. *c*, 5. *f*, 6. *b*
page 6, 1-10:
    Counselor: *read, math, subtract, divide, measure, feet, tools, workers*
    Counselor: *salaries, wage, money, career, parts*

UNIT 2:   SHOP SAFETY

page 8, 2-03: 1. *d*, 2. *c*, 3. *a*, 4. *b*, 5. *d*
page 10, 2-06: *warning, chips, parts, thanks*
page 10, 2-07: *safety glasses, goggles, face shields*
page 11, 2-09: *hands, brush, shop rag, workpiece, change, machine, finished, stop*
page 11, 2-11: *grind, coolant, sharpen, respirator, safe, ventilation*
page 12, 2-13: *50, not, one, help, stock, heavy*
page 13, 2-16: *rags, fire, can, rubber, hot, teacher, extinguisher, blocking*
page 14, 2-18: *clutter, hazard, clean, up, problem, finished, extra, right away*
page 15, 2-20: *wear, caught, things, jewelry, ring, watch, sleeves, hair*
page 16, 2-22: *piece, storage, lights, warning, horseplay, Sorry, air, blowing, skin, running*
pages 16-17, 2-23: 1. *hands, eyes;* 2. *safety glasses;* 3. *goggles;* 4. *short-sleeved, secure;*
    5. *jewelry, gloves;* 6. *ear muffs, ear plugs;* 7. *coolants;* 8. *respirator, ventilated;*
    9. *knees, help;* 10. *chips, hands;* 11. *clutter;* 12. *spills;* 13. *vertical;* 14. *hazards,*
    extinguisher
page 18, 2-24:
    Al: *fire, safety* **Lucy:** *wall* **Le:** *another* **Al:** *more* **Omar:** *work*
    Al: *burn, heat* **Lucy:** *fires* **Al:** *three, paper, flammable, C*
    Al: *rated, cut off* **Lucy:** *tips*

UNIT 3:   MECHANICAL HARDWARE

page 21, 3-03: 1. *T*, 2. *T*, 3. *F*, 4. *F*, 5. *T*, 6. *T*
page 22, 3-06: 1. *Z*, 2. *U*, 3. *S*, 4. *Y*, 5. *Q*, 6. *T*, 7. *L*, 8. *K*, 9. *W*, 10. *G*, 11. *V*,
    12. *Q*, 13. *J*, 14. *J*, 15. *X*
page 23, 3-08: 1. p-*i*- t-c-h   2. r-*o-o* -t   3. *c*- r-*e-s* -t   4. *d-i-a*-m-*e-t-e-r*
    5. t-*h-r*- e-*a-d*   6. *a*-n-g-*l-e*   7. *t-h*-r-*e-a-d* f-*l-a*-t   8. h-*e-l*-i-*x*
    9. m-*i-n-o*-r d-*i-a*-m-*e-t-e-r* 10. *i-n*-t-e-r-i-*o*-r t-*h-r-e*-a-d-*s* 11. m-*a-j*-o-r
    d-*i-a*-m-*e-t-e-r*   12. *e-x*-t-e-r-i-*o*-r t-*h-r-e*-a-d-*s* 13. h-*e-i*-g-*h*-t   14. f-*l-a*-t
page 25, 3-10: 1. *b*, 2. *a*, 3. *b*, 4. *d*, 5. *b*
page 25, 3-11:
    Lucy: *fasteners, tell, exterior* **Lucy:** *hex*   **Omar:** *six, six, nut, threads, bolt*
    Omar: *one, head* **Lucy:** *slots*
page 26, 3-12: 1. *L*, 2. *E*, 3. *M*, 4. *I*, 5. *B*, 6. *F*; 7. *K*, 8. *J*, 9. *D*, 10. *H*, 11. *A*,
    12. *G*, 13. *C*, 14. *L*

page 28, 4-03: 1. *d*, 2. *b*, 3. *d*
page 31, 4-05: 1. *b*, 2. *d*, 3. *b*, 4. *c*, 5. *d*, 6. *a*, 7. *a*
page 31, 4-06:

Several kinds of drawings (d's): *exploded view* d's, *isometric* d's and *perspective* d's. One most often used in the shop = *orthographic* projection. Shows 2 or 3 *views* . Enough to visualize part. 3 usual *views* = front, *top*, & *right* side. Iso*metric* d's look more *real*, because they show more than one *view* of part at one time. Leading corner is made with 3 angles of *120°*. Difference between these and per*spective* d's is that in a per*spective* d, farther into picture you go, the *smaller* things look, but in an iso*metric* d all *units* are the same. For next time study pp. *26-35* in TB. Fri. quiz on *nomenclature* of lines. (nomenclature = technical names).

page 36, 4-09: 1. *H*, 2. *D*, 3. *G*, 4. *B*, 5. *C*, 6. *F*, 7. *A*, 8. *I*, 9. *E*
page 37, 4-10:

1. *Why do center lines come in pairs?*   2. *When do you use short break lines?*
3. *Where are dimension lines drawn?*   4. *Who uses these lines to communicate?*
5. *How do I show the inside of some part?*   6. *When do I use leader lines?*   7. *Which lines are used to show the main outlines of the part?*   8. *What kind of line shows the length of a part's edge?*   9. *What's the shape of a round stock break line?*

page 40, 4-13:

1. *C, D, E*   2. *.0005*, upper limit = *2.7535*, lower limit = *2.7525*
3. *16, 63, the bearing*   4. *C*, 5. *A*, 6. *B*, 7. *B*, 8. *A*, 9. *C*

UNIT 5: HAND TOOLS

page 43, 5-03: 1. *b*, 2. *c*, 3. *d*, 4. *b*, 5. *a*, 6. *c*
page 43, 5-04: 1. *C*, 2. *B*, 3. *A*, 4. *B*
page 45, 5-07: 1. *T*, 2. *F*, 3. *T*, 4. *T*, 5. *F*, 6. *F*
page 47, 5-09: 1. *d*, 2. *j*, 3. *f*, 4. *b*, 5. *i*, 6. *g*, 7. *h*, 8. *a*, 9. *e*, 10. *c*
page 48, 5-12:

1. *threads*   *rethreading, diestock*
2. *external*   *collet, diestock*
3. *blind*   *taps, tap*
4. *tolerance, hole*   *reamer*
5. *keyseat*   *square, sizes*
6. *filing*   *tang, forward, drag*

page 50, 5-14: 1. *G*, 2. *E*, 3. *H*, 4. *B*, 5. *A*, 6. *F*, 7. *D*, 8. *C*, 9. *K*, 10. *I*, 11. *M*, 12. *J*, 13. *O*, 14. *N*, 15. *L*, 16. *S*, 17. *P*, 18. *Q*, 19. *R*, 20. *W*, 21. *V*, 22. *T*, 23. *U*, 24. *AA*, 25. *Z*, 26. *CC*, 27. *X*, 28. *BB*, 29. *Y*, 30. *EE*, 31. *GG*, 32. *JJ*, 33. *HH*, 34. *FF*, 35. *DD*, 36. *II*

UNIT 6: THE LANGUAGE OF MATHEMATICS

page 52, 6-03:   m. *473*   n. *2,350,043*   o. *75,000*   p. *2,007*   q. *65,302*
r. *6,800,512*   s. *9,185*   t. *78*
page 53, 6-05:   p. *4/5*   q. *4 4/50*   r. *17/32*   s. *60/64*   t. *102 3/7*   u. *8 125/1000*
v. *3/10000*   w. *75 75/100*   x. *23 7/16*   y. *49 5/8*
page 54, 6-07:   e. *59.35*   f. *28.05*   g. *.875*   h. *3.125*   i. *.0001*

UNIT 6: THE LANGUAGE OF MATHEMATICS (cont.)   ANSWERS

page 54, 6-08:   k. *.12*   l. *.125*   m. *9.032*   n. *3.1416*   o. *2.2*   p. *819.019*   q. *3.14*
r. *822.750*   s. *12.500*   t. *92.3756*

page 55, 6-10:   1. *T*, 2. *F*, 3. *T*, 4. *F*

page 56, 6-11:
Lucy: *problems* Omar: *practice* Lucy: *multiplication*
Omar: *product, division* Lucy: *quotient, numbers*
Omar: *dividend, divisor, head* Lucy: *sum, 9, difference, 17*
Omar: *addition, 24, difference, 13*

page 56, 6-12:   **1.** a. *.05*   b. *.125*   c. *.0075*   d. *42.35*   e. *7.9*   f. *.0185*
2. a. *1/2*   b. *3/4*   c. *8 1/8*   d. *35 6/7*   e. *2/3*   f. *5 17/20*
3. a. *12/100 = 3/25*   b. *4/20/1000 = 4 1/50*   c. *117/1000*   d. *12 500/1000 = 12 1/2*
e. *25 4/100 = 25 1/25*   f. *52 625/1000 = 52 5/8*

page 58, 6-15:   1. *F*, 2. *T*, 3. *T*, 4. *F*

page 59, 6-17:   1. *T*, 2. *T*, 3. *F*, 4. *T*

page 60, 6-18:   1. *h*, 2. *e*, 3. *j*, 4. *i*, 5. *g*, 6. *b*, 7. *d*, 8. *f*, 9. *a*, 10. *c*

page 60, 6-19:   1. *B*, 2. *B*, 3. *A*, 4. *C*, 5. *C*, 6. *A*, 7. *C*, 8. *A*, 9. *B*

page 62, 6-21:   **1.** *hypotenuse*, **2.** *opposite*, **3.** *adjacent*, **4.** *adjacent*, **5.** *opposite*,
6. $c^2 = a^2 + b^2$, **7.** a. *9*, b. *4.41*, c. *49*, d. *28.09*, **8.** a. *9*, b. *10*, c. *2*, d. *1/2*

page 62, 6-22:   *25.5 cm, 7.85 cm, 3.7 cm, 3/4 in. dia., 42 mm, 25 mm, I.D., 6 mm,
max. dia., 6.5 mm, min. dia., 6 mm, 21 TPI, 45° cham*

page 66, 6-27:   **LECTURE NOTES**

right-triangle, formulas, 798, trigonometry

angle A	side a	side b	side c
$\underline{sin\ A}\ =\ \underline{a}$ c	a = c x $\underline{sin\ A}$		$\underline{c}\ =\ \underline{a}$ sin A
$\underline{cos}$ A = $\underline{b}$ c		b = c x $\underline{cos\ A}$	c = $\underline{b}$ cos A
$\underline{tan\ A}\ =\ \underline{a}$ b	a = $\underline{b}$ x tan A	b = $\underline{\ \ \ \ a\ \ \ \ }$ $\underline{tan\ A}$	

formula, 1. *c = 12, a = 6.883*   2. *a = 7.125, sin A = 5.000, c = 14.250*
3. *a = 6, c = 10, theorem, $b^2 = 64$, b = 8*

pages 67-68, 6-28:   **Mathematics Terminology Test**

A.:   **1.** a. *305*   b. *50*   c. *1,322,514*   d. *15*   e. *142,860*   f. *16*
2. a. *1/2*   b. *5/32*   c. *13/64*   d. *5 3/4*   e. *105 11/16*   f. *53/4*
3. a. *0.5*   b. *4.15*   c. *18.003*   d. *8.003*   e. *0.500*   f. *0.875*   g. *3.1416*
h. *2945.120*   i. *18.080*
4. a. *24-0176-7Z*   b. *PDQ-457-T34*   c. *99-0573-4812*
5. a. *45°*   b. *13° 14′ 12″*   c. *120° 0′ 15″*

UNIT 6: THE LANGUAGE OF MATHEMATICS (cont.) ANSWERS

pages 67-68, 6-28:   Mathematics Terminology Test

B.: 1. *operations*, 2. *reducing*, 3. *product*, 4. *difference*, 5. *numerator*, 6.*denominator*,
7. *divisor*, 8. *dividend*, 9. *I = 180°*, *II. = 180°*, *III. = 180°*, 10. *right triangle*,
11. *hypotenuse*, 12. *adjacent side*, 13. *sine*, 14. a. *circumference*, b. *radius*,
c. *diameter*, d. *center*, e. *chord*, 15. *pi*

C.: 1. a. inch = *in.* b. foot = *ft.* c. 25 degrees = *25°* d. 4 centimeters = *4 cm*
e. 25.3 millimeters = *25.3 mm* f. the sine of angle B = *sin B*
g. the cosine of angle A = *cos A* h. the tangent of angle B = *tan B*
2. a. dia. = *diameter* b. ID = *inside diameter* c. min. = *minimum*
d. CS = *cutting speed* e. dim. = *dimension* f. max. = *maximum*
g. RPM = *revolutions per minute* h. OD = *outside diameter*
3. a. $C = d\pi$ b. $RPM = \dfrac{4CS}{D}$ c. $c^2 = a^2 + b^2$

UNIT 7: MEASURING INSTRUMENTS

page 70, 7-03: 1. *Does the USA mass produce many parts?* 2. *Did Beatriz measure the part carefully?* 3. *Did Al teach the class about accuracy?* 4. *Do the students understand discrimination?* 5. *Do Germany and Japan have quality standards?* 6. *Did the machinist choose his tools for the job?*

page 71, 7-05: *metrology, calibrated, accurate, precision, reliable, ten, discriminate*

page 73, 7-07:   Part A: Dimensions
Figure 7-01

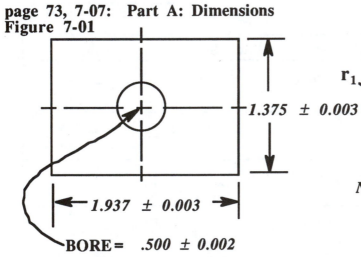

Figure 7-02

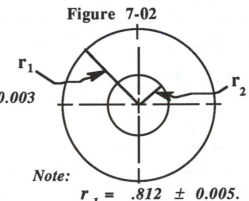

Note:
$r_1 = .812 \pm 0.005.$
$r_2 = 312 \pm 0.001$

Part B: Decimal Equivalents
    often, equivalents, memorize, decimal, decimal equivalents, calculator
Part C: Conversation:
    fraction, equivalent, 1/64, .375, .062, 1/64, .016

1.	1/2	=	.500
2.	1/4	=	.250
3.	3/4	=	.750
4.	1/8	=	.125
5.	3/8	=	.375
6.	5/8	=	.625
7.	7/8	=	.875
8.	1/16	≈	.062
9.	1/32	≈	.031
10.	1/64	≈	.016

page 74, 7-09: *measurement, inch, nations, SI, metric, centimeter, fingernail, millimeter, dime,* 1 in. = 25.4 mm, 1 in. = 2.54 cm, 1 mm = .03937

page 75, 7-11   ❶ *9/32*   ❷ *11/16*   ❸ *27/64*   ❹ *29/32*

page 78, 7-14: 1. a. *inside micrometer,* b. *outside micrometer,* c. *depth micrometer,* 2. *.001 inches,* 3. *.0001 inches,* 4. *extension rods*

page 78, 7-16:

		A.				B		
			1	*.200*			1	*.400*
			2	*.050*			2	*.075*
			3	*.014*			3	*.018*
			Total	= *.264*			Total	= *.493*

page 79, 7-18:

		A.				B		
			1	*.400*			1	*.100*
			2	*.075*			2	*.025*
			3	*.003*			3	*.007*
			4	*.0002*			4	*.0004*
			Total	= *.4782*			Total	= *.1324*

page 80, 7-19: 1. *B,* 2. *C,* 3. *C,* 4. *A,* 5. *A,* 6. *C,* 7. *B*

page 80, 7-20: 1. *anvil,* 2. *spindle,* 3. *locking nut,* 4. *sleeve,* 5. *thimble,* 6. *ratchet,* 7. *frame,* 8. *outside micrometer,* 9. *inside micrometer,* 10. *sleeve,* 11. *extension rods,* 12. *depth micrometer*

page 83, 7-24: 1. *inside jaws,* 2. *main scale,* 3. *bar* or *beam,* 4. *vernier caliper,* 5. *outside calipers,* 6. *caliper legs,* 7. *adjusting nut,* 8. *vernier scale,* 9. *clamping screw,* 10. *adjusting nut,* 11. *dial,* 12. *thumb pad,* 13. *depth gage blade,* 14. *dial calipers,* 15. *movable jaws*

page 83, 7-25: 1. *Do the vernier scales slide along the main scale?* 2. *Did you lock the reading in place with the clamping screw?* 3. *Do outside calipers have two legs and an adjusting nut?* 4. *Did Danny insert the depth gage blade into the hole?* 5. *Do (or did) the anvil and the spindle fit snugly around the part?* 6. *Does Carol use extension rods on her depth micrometer?*

page 84, 7-27: 1. *cylindrical plug gages,* 2. *screw pitch gage,* 3. *telescoping gages,* 4. *thread plug gage,* 5. *ring gages,* 6. *small hole gages,* 7.
8.    9. 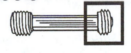

10.   11. (Draw whatever instrument you want.)

UNIT 8: LAYOUT TOOLS & PROCEDURES

page 88, 8-05: 1. *B,* 2. *A,* 3. *C,* 4. *C,* 5. *B,* 6. *A*

page 89, 8-06: 1. *Study the drawing on page 247.* 2. *Cut off a 6 1/16 in. piece of rough stock.* 3. *Lay down a paper towel.* 4. *Apply layout dye to the workpiece.* 5. *Scribe the major reference lines.* 6. *Measure & scribe the angles and hole centers.* 7. *Pin punch the hole centers.* 8. *Center punch the hole centers.*

page 89, 8-07: 1. *combination square, rule, and scriber*
2. *Figure E-37, page 247*
3. *dividers*
4. *Fig. E-47 & Fig. E-48*

5.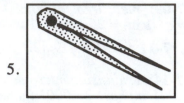

page 89, 8-08: 1. p-*i*-n p-*u*-n-c-*h* 2. s-*t*-e-e-*l* *r*-u-*l*-e 3. l-*a*-y-o-*u*-t d-*y*-e
4. s-*q*-u-a-*r*-e h-*e*-a-d 5. s-*c*-r-*i*-b-e-*r* 6. *l*-e-*v*-e-*l* 7. l-a-y-o-*u*-t h-*a*-m-*m*-e-*r*
8. *b*-e-*v*-e-*l* p-*r*-o-*t*-r-a-*c*-t-o-*r* 9. *c*-o-m-*b*-i-n-*a*-t-i-o-*n* s-e-*t* 10. *c*-e-n-*t*-e-*r*
p-*u*-n-*c*-h 11. *s*-*u*-*r*-*f*-a-*c*-e p-*l*-a-*t*-e 12. c-e-n-*t*-e-*r* h-*e*-a-d

page 92, 8-12: 1. *Place workpiece on edge.* 2. *Use reference surface...* 3. *Scribe .750 for frame width...* 4. *Place workpiece on its side...* 5. *Place the workpiece on end clamped to an angle plate...*

page 94, 8-14: A. *dial indicator*, B. *surface plate*, C. *surface gage*, D. *layout dye*, E. *center head*, F. *bevel protractor*, G. *square head*, H. *vernier height gage*, I. *gage block set*, J. *parallels*, K. *angle plate*, L. *prick punch*, M. *center punch*, N. *sine bar*, O. *layout hammer*, P. *scriber*
1. *P*, 2. *A*, 3. *I*, 4. *B*, 5. *K*, 6. *F*, 7. *M*, 8. *G*

UNIT 9: SAWING MACHINES

pages 97-98, 9-03: **SECTION TITLE:** *G: Sawing Machines*
A. 1. *cutoff machines bar stock tools*
2. *vertical band saw* a. *shaped size cuts* b. *curved*
B. 1. *Reciprocating* b. *column type*
2. *Horizontal Cutoff Machine* Note: *blade teeth edge*
3. *Universal Tilt Cutoff*
4. *Abrasive Cutoff Machine* Note: *nonmetallic glass brick stone*
5. *Cold Saw Cutoff Machines circular*
C. 1. *Band Machine with Fixed Worktable* Notes: b. *tilted*
2. Note: *worktable operator push*
3. *High Velocity* Note: *10 15,000*
4. *Capacity* Notes: *large stationary moved*

page 98, 9-04: 1. *T*, 2. *F*, 3. *T*, 4. *T*, 5. *T*, 6. *F*, 7. *T*

page 99-100, 9-05: 1. *cutting oils soluble cutting fluid* 2. *fluid water based coolant* 3. *removal workpiece small finishing roughing* 4. *width cut kerf kerf kerf* 5. *bottom teeth gullet* 6. *angle keenness rake rake rake* 7. *angle clearance relief angle* 8. *teeth inch pitch width* 9. *wider set gage* 10. *metals cooled hardness to anneal*

page 100, 9-06: 1. *T*, 2. *T*, 3. *F*, 4. *T*, 5. *T*, 6. *T*, 7. *F*

page 101, 9-08:
Lucy: *cutoffs* **Le:** *bar stock* **Lucy:** *vertical rough cut pitch* **Le:** *12 harder*

page 102, 9-09: 1. *B*, 2. *O*, 3. *K*, 4. *F*, 5. *A*, 6. *N*, 7. *C*, 8. *J*, 9. *M*, 10. *I*, 11. *G*, 12. *P*, 13. *E*, 14. *L*, 15. *H*, 16. *D*

page 104, 10-02:
 S: *sensitive* **I:** *red, green, start, stop* **S:** *controls* **I:** *two*
 S: *automatic, manual* **I:** *manual, feed, pressure* **S:** *manual controls*

page 106, 10-04: 1. *F*, 2. *T*, 3. *T*, 4. *T*, 5. *T*, 6. *T*, 7. *T*

page 107, 10-06: 1. *F*, 2. *T*, 3. *F*, 4. *T*, 5. *T*

page 108, 10-07: A. s-*p*-e-e-*d* c-o-n-*t*-*r*-o-l   B. O-*n*-/O-*f*-*f* b-*u*-t-*t*-o-*n*-s
 C. *d*-e-*p*-t-*h* s-*t*-o-*p*   D. s-*p*-i-n-*d*-d-l-*e*   E. *t*-a-*b*-l-e   F. *c*-o-l-u-*m*-*n*
 G. *b*-a-s-e   H. s-*e*-n-s-*i*-*t*-*i*-*v*-e *d*-r-*i*-l-l *p*-*r*-*e*-s-s   I. t-*a*-*b*-l-e l-o-c-*k*
 J. *c*-h-*u*-c-k   K. h-*a*-n-*d* *f*-e-e-*d* l-e-*v*-e-*r*   L. *m*-o-*t*-o-*r*
 M. d-*r*-i-*l*-l-i-*n*-*g* h-*e*-a-d   N. r-*a*-d-*i*-a-*l* a-*r*-*m*   O. *a*-*r*-*m* e-*l*-e-*v*-a-t-*i*-*n*-g
 l-e-*v*-e-*r* d-*y*-e   P. r-*a*-d-*i*-a-*l* *a*-*r*-*m* *d*-*r*-*i*-l-l p-*r*-*e*-s-s

pages 108-9, 10-08: 1. *F*, 2. *G*, 3. *B*, 4. *C*, 5. *I*, 6. *K*, 7. *N*, 8. *E*, 9. *M*, 10. *J*,
 11. *D*, 12. *P*, 13. *L*, 14. *A*, 15. *H*, 16. *O*

page 111, 10-11: 1. *B*, 2. *C*, 3. *A*, 4. *B*, 5. *A*, 6. *B*, 7. *C*

page 113, 10-14: 1. *K*, 2. *I*, 3. *D*, 4. *L*, 5. *A*, 6. *B*, 7. *J*, 8. *E*, 9. *H*,
 10. *G*, 11. *C*, 12. *F*

page 114, 10-16: (1) *shank*, (2) *body*, (3) *point*, *axis*, *center*, *flutes*, *one*, *chips*,
 hole, *cutting fluid*, *helix angle*

page 115, 10-18: 1. *E*, 2. *O*, 3. *M*, 4. *P*, 5. *S*, 6. *N*, 7. *U*, 8. *T*, 9. *R*, 10. *O*, 11. *Q*

page 116, 10-20: *diameter, micrometer, margins, chisel edge, cutting lips, cutting, hole,*
 clearance, flutes, angle, point, 118°, 135°

page 117, 10-22:

Decimal inch	Drill size	Decimal inch	Drill size
.0313 in.	1/32	.0595 in.	*53*
.0394 in.	1 mm	.2420 in.	*C*
.0400 in.	60	.2638 in.	*6.7 mm*
.0550 in.	*54*	.3906 in.	*15/64*
.2420 in.	*1/8*	.4040 in.	*Y*

page 118, 10-23: 1. *C*, 2. *E*, 3. *BB*, 4. *Y*, 5. *D*, 6. *AA*, 7. *DD*, 8. *B*, 9. *X*, 10. *A*,
 11. *CC*, 12. *W*, 13. *Z*, 14. *F*, 15. *S*, 16. *V* 17. *J*, 18. *K*, 19. *U*, 20. *H*, 21. *R*,
 22. *I*, 23. *G*, 24. *Q*, 25. *P*, 26. *L*, 27. *N*, 28. *T*, 29. *M*, 30. *O*

UNIT 11: TURNING MACHINES

page 120, 11-03: 1. *T*, 2. *T*, 3. *T*, 4. *F*, 5. *F*

page 122-123, 11-05:
 Al: *engine main bed* **Lucy:** *bed piece top supports* **Al:** *headstock*
 Le: *left-hand spindle workpiece rotation hold free* **Al:** *turns* **Le:** *gears*
 Al: *review tailstock* **Omar:** *ways right-hand spindle tip rotate* **Al:** *ways*
 Lucy: *carriage toolbit fits toolpost direction tangle* **Al:** *compound rest*
 Le: *slide sits right operator part rotating motion handwheel* **Le:** *feed*
 Omar: *base level quick-change gearbox power* **Al:** 1) *headstock;* 2) *tailstock;*
 3) *bed;* 4) *carriage;* 5) *quick-change gearbox;* and 6) *base*

page 124, 11-06: 1. *b*, 2. *d*, 3. *c*, 4. *a*, 5. *c*, 6. *b*, 7. *c*, 8. *a*, 9. *b*, 10. *d*

page 125, 11-07: 1. *L*, 2. *R*, 3. *P*, 4. *E*, 5. *I*, 6. *J*, 7. *A*, 8. *Q*, 9. *N*, 10. *U*, 11. *O*,
 12. *H*, 13. *K*, 14. *C*, 15. *S*, 16. *G* 17. *T*, 18. *F*, 19. *M*, 20. *B*, 21. *D*

page 129, 11-11:
1. **S:** *turning rough* **T:** *soft rake 12"* 2. **S:** *scratches* **T:** *edge radius 1/16* 3. **S:** *chattering* **T:** *smaller* 4. **S:** *tools select advice* **T:** *high speed carbide* 5. **S:** *chip* **T:** *grind side curl break nine*

page 130, 11-12: 1. *F,* 2. *T,* 3. *F,* 4. *F,* 5. *T*

page 130, 11-13: 1. *B,* 2. *I,* 3. *D,* 4. *G,* 5. *C* 6. *A,* 7. *H,* 8. *E,* 9. *J,* 10. *F,* 11. *K*

page 133, 11-16: 1. *left-hand facing* 2. *cut-off* 3. *external radius* 4. *external threads* 5. *right-hand facing* 6. *left-hand roughing* 7. *right-hand roughing* 8. *recess* 9. *fillet radius*

page 134, 11-17: 1. *C,* 2. *A,* 3. *B,* 4. *C,* 5. *A*

page 134, 11-18:
1. A. *r.h. facing tool* B. *r.h. roughing tool* C. *cut-off tool*
2. A. *l.h. facing tool* B. *fillet radius tool* C. *cut-off tool*
3. A. *fillet radius tool* B. *external radius tool* C. *r.h. roughing tool*
4. A. *l.h. roughing tool* B. *recessing tool* C. *parting tool*
5. A. *recessing tool* B. *threading tool* C. *r.h. facing tool*

page 137, 11-21: 1. *F,* 2. *T,* 3. *T,* 4. *F,* 5. *T,* 6. *F,* 7. *T*

page 139, 11-25: 1. *F,* 2. *F,* 3. *T,* 4. *T,* 5. *F,* 6. *T,* 7. *T,* 8. *F*

page 140, 11-27:
Al: *devices lathe* **Lucy:** *plates dogs between* **Omar:** *independent universal round collet* **Lucy:** *tapered mandrel*

pages 141-142, 11-28:
Part A: 1. *F,* 2. *A,* 3. *D,* 4. *C,* 5. *E,* 6. *B,* 7. *S,* 8. *J,* 9. *O,* 10. *L,* 11. *I,* 12. *M,* 13. *P,* 14. *U,* 15. *H,* 16. *R* 17. *G,* 18. *Q,* 19. *N,* 20. *K,* 21. *T*
Part B: 22. *W,* 23. *CC,* 24. *V,* 25. *FF,* 26. *AA,* 27. *EE,* 28. *BB,* 29. *Z,* 30. *X,* 31. *DD,* 32. *Y*
Part C: 33. *KK,* 34. *II,* 35. *MM,* 36. *HH,* 37. *NN,* 38. *JJ,* 39. *LL,* 40. *GG*
Part D: 41. *ZZ,* 42. *OO,* 43. *XX,* 44. *QQ,* 45. *UU,* 46. *RR,* 47. *BBB,* 48. *YY,* 49. *VV,* 50. *AAA,* 51. *PP,* 52. *TT,* 53. *WW,* 54. *SS*

UNIT 12: *VERTICAL MILLING MACHINES*

pages 144-45, 12-03: 1. *d,* 2. *a,* 3. *c,* 4. *b,* 5. *c,* 6. *b,* 7. *c,* 8. *a,* 9. *b*

page 147, 12-06: 1. *base & column,* 2. *knee,* 3. *saddle,* 4. *table,* 5. *ram,* 6. *toolhead*

page 147-48, 12-08: 1. *T,* 2. *T,* 3. *F,* 4. *F,* 5. *T,* 6. *F*

page 148, 12-10: 1. *F,* 2. *T,* 3. *T,* 4. *F,* 5. *T,* 6. *F*

page 149, 12-12: 1. *T,* 2. *F,* 3. *F,* 4. *T,* 5. *T,* 6. *F,* 7. *T*

page 150, 12-13: 1. *C,* 2. *A,* 3. *B,* 4. *A,* 5. *C,* 6. *B*

page 150, 12-14: 1. *B,* 2. *A,* 3. *A,* 4. *C,* 5. *B,* 6. *C,* 7. *C,* 8. *B,* 9. *A,* 10. *C*

page 151, 12-15: 1. *F,* 2. *A,* 3. *D,* 4. *N,* 5. *I,* 6. *J,* 7. *M,* 8. *E,* 9. *G,* 10. *C,* 11. *H,* 12. *B,* 13. *L,* 14. *K*

page 152, 12-16: 1. *N,* 2. *A,* 3. *G,* 4. *F,* 5. *J,* 6. *C,* 7. *K,* 8. *E,* 9. *L,* 10. *I* 11. *B,* 12. *M,* 13. *D,* 14. *H*

page 153, 12-18: 1. *periphery,* 2. *flute,* 3. *dovetail,* 4. *T-slots, T-nuts,* 5. *hog, roughing* 6. *keyseat*

UNIT 12: VERTICAL MILLING MACHINES (cont.) ANSWERS

page 156, 12-20: 1. *a,* 2. *d,* 3. *c,* 4. *c,* 5. *d,* 6. *b,* 7. *c,* 8. *b*

page 158, 12-23: 1. *C,* 2. *B,* 3. *C,* 4. *B,* 5. *A,* 6. *A*

page 159, 12-25: 1. *B,* 2. *H,* 3. *G,* 4. *D,* 5. *J,* 6. *C,* 7. *I,* 8. *A,* 9. *K,* 10. *F,* 11. *E,* 12. *J*

page 161, 12-28: 1. *A,* 2. *A,* 3. *C,* 4. *B,* 5. *B,* 6. *C*

page 161, 12-29: *care, aligned, angular, loosening, retighten, all, at one time, swivels, pressure, weight, upside, injury, damage, two*

page 162, 12-30: **Names:** 1. *K,* 2. *A,* 3. *G,* 4. *I,* 5. *C,* 6. *J,* 7. *B,* 8. *E,* 9. *H,* 10. *D,* 11. *G,* 12. *F;* **Uses:** 13. *G,* 14. *C,* 15. *H,* 16. *A,* 17. *K,* 18. *F,* 19. *I,* 20. *D,* 21. *B,* 22. *E,* 23. *J*

UNIT 13: HORIZONTAL MILLING MACHINES

page 165, 13-04: 1. *T,* 2. *T,* 3. *F,* 4. *T,* 5. *F,* 6. *T,* 7. *F*

page 166, 13-08: 1. *F,* 2. *T,* 3. *T,* 4. *F,* 5. *T*

page 169, 13-11: 1. *T,* 2. *F,* 3. *T,* 4. *F,* 5. *T,* 6. *T,* 7. *T,* 8. *F*

page 170, 13-13:

Lucy: *That sounds like a stagger tooth side milling cutter.*

Omar: *The slitting saw is for cutting slots and for cutting lengths off of rough stock.*

Le: *You could use a convex milling cutter.*

page 171, 13-15: 1. *T,* 2. *F,* 3. *T,* 4. *T,* 5. *T*

page 171-172, 13-16: 1. *F,* 2. *B,* 3. *A,* 4. *G,* 5. *C,* 6. *E,* 7. *D,* 8. *M,* 9. *I,* 10. *P,* 11. *R,* 12. *K,* 13. *O,* 14. *H,* 15. *L,* 16. *Q,* 17. *N,* 18. *J,* 19. *W,* 20. *BB,* 21. *T,* 22. *X,* 23. *Z,* 24. *EE,* 25. *V,* 26. *CC,* 27. *AA,* 28. *Y,* 29. *S,* 30. *U,* 31. *DD,* 32. *JJ,* 33. *OO,* 34. *KK,* 35. *FF,* 36. *II,* 37. *NN,* 38. *LL,* 39. *HH,* 40. *MM,* 41. *GG*

UNIT 14: GRINDING MACHINES

page 174, 14-03: 1. *h,* 2. *g,* 3. *e,* 4. *a,* 5. *f,* 6. *d,* 7. *b,* 8. *c*

page 175, 14-05: a. *fused, oxide, alloys, iron* b. *ceramic, aluminum, rough, wheel* c. *silicon, friable* d. *boron nitride, ferrous* e. *diamond, manufactured, carbide*

page 175, 14-06: 1. *d,* 2. *f,* 3. *e,* 4. *c,* 5. *a,* 6. *b*

page 177, 14-09: 1. *F,* 2. *F,* 3. *F,* 4. *T,* 5. *T,* 6. *F,* 7. *T,* 8. *T*

pages 177-178, 14-10:

Al: *talked, safety, study, grinding, rules, general, glasses, shields, swarf location, controls, manufacturers, arrangement, different, switches, motor dress, moving, hair, jewelry, rings coolant, spills*

Al: *how, wheel, easily, pieces, cracks, end, tap, cracked, not*

pages 178-179, 14-11: *Compare your drawings and labeling with those given on page 649 of the textbook. Put a check here for each correct picture:*

Type 1 ☐ Type 2 ☐ Type 6 ☐ Type 11 ☐ Type 12 ☐

page 180, 14-14:

T: *procedure* **S:** *first* *spindle* *Second* *point* *touches* *Finally* *dresser*

T: *truing* **S:** *more* *concentric* *reshaping*

page 182, 14-17: 1. *F,* 2. *T,* 3. *F,* 4. *T,* 5. *T,* 6. *T*

page 185, 14-22: 1. *c,* 2. *a,* 3. *b,* 4. *d,* 5. *c*
page 185, 14-23: 1. *G,* 2. *I,* 3. *N,* 4. *O,* 5. *A,* 6. *M,* 7. *B,* 8. *H,* 9. *D,* 10. *K,*
   11. *E,* 12. *L,* 13. *J,* 14. *C,* 15. *F*
page 188, 14-27: 1. *d,* 2. *d,* 3. *d,* 4. *a*
page 188, 14-28: 1. *Q,* 2. *I,* 3. *K,* 4. *E,* 5. *M,* 6. *G,* 7. *O,* 8. *B,* 9. *N,* 10. *A,*
   11. *J,* 12. *R,* 13. *L,* 14. *F,* 15. *P,* 16. *D,* 17. *H,* 18. *C*

UNIT 15: CNC MACHINING

page 191, 15-03: Any of the following describe recent advances:
   *1. The use of CNC machines is increasing.   2. Computers and computer programs are
   controlling the making of parts. 3. The quality of tools is continually improving. 4. More
   non-traditional tools are being used. 5. Greater productivity is possible with CNC.*
page 192, 15-06: 1. *T,* 2. *F,* 3. *T,* 4. *F,* 5. *F*
page 194, 15-10: 1. *operator,* 2. *programmer,* 3. *operator,* 4. *setup person,*
   5. *setup person,* 6. *programmer*

page 196, 15-13:

Name	X	Y	Quadrant
A	3.000	-5.000	IV
B	-2.000	2.000	II
C	2.000	4.000	I
D	-5.000	0.000	on axis
E	4.500	1.500	I
F	-1.250	-3.250	III
G	4.750	-2.500	IV
H	-3.875	-2.125	III
I	-4.000	4.000	II
J	1.875	-4.500	IV

page 196, 15-14: 1. *T,* 2. *T,* 3. *F,* 4. *T,* 5. *F,* 6. *T,* 7. *F,* 8. *T*
page 198, 15-17: 1. *C,* 2. *C,* 3. *A,* 4. *B,* 5. *A,* 6. *B*
page 198, 15-18: 1. *H,* 2. *A,* 3. *G,* 4. *F,* 5. *C,* 6. *D,* 7. *E,* 8. *B,* 9. *I*

STUDY GUIDE GLOSSARY

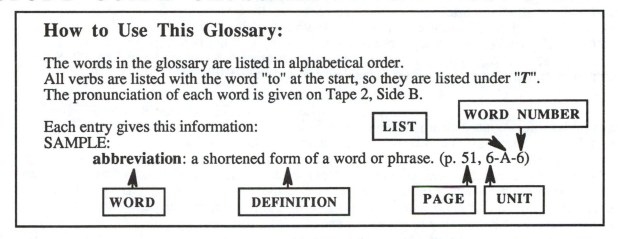

How to Use This Glossary:

The words in the glossary are listed in alphabetical order.
All verbs are listed with the word "to" at the start, so they are listed under "*T*".
The pronunciation of each word is given on Tape 2, Side B.

Each entry gives this information:
SAMPLE:

LIST

WORD NUMBER

abbreviation: a shortened form of a word or phrase. (p. 51, 6-A-6)

WORD   DEFINITION   PAGE   UNIT

A

abbreviation: a shortened form of a word or phrase. (p. 51, 6-A-6)
abrasive: a substance, such as finely ground aluminum oxide, used for grinding, smoothing, or polishing. (p. 95, 9-A-8)
abrasive grain: a tiny, hard piece of abrasive material. (p. 173, 14-A-1)
absent: not being at work or class. (p. 7, 2-A-7)
absolute: unchanged by other changes within a system. (p. 197, 15-D-4)
acceptable: meeting a certain standard that was set ahead of time. (p. 38, 4-C-3)
accessory: a piece of equipment added to help something work better. (p. 90, 8-B-3)
accident: being hurt when it is not expected. (p. 7, 2-A-3)
accuracy: a condition of being correct, especially with measurements. (p. 4, 1-B-7)
adapter: a part that helps one part fit into a holder of a different size. (p. 143, 12-A-5)
adaptive: able to change to take care of new or developing situations. (p. 191, 15-B-5)
adjacent: located next to something else. (p. 57, 6-B-4)
adjustable: able to be changed so it will fit. (p. 44, 5-B-1)
adjustment: making a machine accurate by changing the controls. (p. 109, 10-B-9)
advance: making something better with more efficiency or with less cost. (p. 189, 15-A-2)
advice: an opinion, given to another person, about a particular situation. (p. 126, 11-B-1)
alloy: a fused mixture of two or more metals or a metal and some chemicals. (p. 173, 14-A-7)
alternate: placed first on one side then on the other; every other one. (p. 167, 13-B-5)
arrangement: the way a group of things are laid out for use or for appearance. (p. 167, 13-B-2)
assembly: a fitting together of parts to make a whole. (p. 19, 3-A-2)
attitude: a mental outlook on some situation. (p. 7, 2-A-8)
automated: self-operated by means of computer programming. (p. 119, 11-A-5)
automatic: done by the power of a machine motor. (p. 104, 10-A-10)
average: the number you get by adding two or more quantities and dividing by the number of quantities. (p. 19, 3-A-10)
axis: an invisible line that passes through the center of a round object. (p. 109, 10-B-10)

B

bandsaw blade: a continuous loop of steel with cutting teeth on one edge. (p. 95, 9-A-4)
base: the part on which something rests. (p. 103, 10-A-3)
beam (bar): the straight rod upon which the movable jaws slide and which shows the graduation marks of the main scale. (p. 76, 7-B-8)
bearing: a supporting part that carries the weight of another part. (p. 38, 4-C-9)

bevel: an angle other than a 90° angle. (p. 85, 8-A-9)

bilateral: having two sides or two measurements. (p. 38, 4-C-4)

blank: the small bar of metal used to make a toolbit. (p. 126, 11-B-3)

brittle: easily broken or shattered because it is hard and will not bend. (p. 176, 14-B-2)

build-up: several gage blocks stacked on top of each other to a desired height. (p. 90, 8-B-9)

burr: a rough or sharp edge left on a metal part by cutting or drilling. (p. 44, 5-B-6)

C

caliper (calipers): an instrument with a pair of jaws used to measure lengths. (p. 76, 7-B-5)

capacity: the amount that can be received or held. (p. 103, 10-A-2)

career: a field of work that a person trains for and continues in for a long time. (p. 4, 1-B-1)

Cartesian coordinates: a pair of numbers that locate a point by its distances from two intersecting perpendicular lines in the same plane, the X axis and the Y axis. (p. 195, 15-C-3)

cast iron: hard iron which has been melted and formed in a desired shape. (p. 119, 11-A-7)

ceramic: made of baked clay or glass-like material from the earth. (p. 173, 14-A-8)

chamfer: a cut made along the edge of a part to remove the sharpness of the edge. (p. 32, 4-B-9)

chips: small pieces of metal that are cut away from the workpiece. (p.1, 1-A-5)

chisel: a sharp-edged tool for cutting or shaping wood, stone, or metal. (p. 111, 10-C-3)

chuck: a metal clamping device used to hold a rotating tool or rotating work. (p. 103, 10-A-5)

clearance (1): the clear space between moving objects or parts. (p. 109, 10-B-8)

clearance (2): grinding an angle behind any cutting edge so that the material being cut can flow more easily past the cutting edge. (p. 126, 11-B-6)

cluster: a number of things of the same kind gathered together. (p. 179, 14-C-3)

CNC: *computerized numerical control;* computer controls machine's making a part. (p.1, 1-A-8)

coarse: made up of larger sizes. (p. 23, 3-B-4)

code: a set of symbols or words used to control the work of a computer. (p. 191, 15-B-7)

column: a slender, cylindrical shaft strong enough to support heavy objects. (p. 103, 10-A-4)

complex: made up of two or more interrelated parts. (p. 191, 15-B-1)

compressed air: air which is held under pressure. (p. 12, 2-C-4)

concave: curving inward like the inside surface of half a rubber ball. (p. 131, 11-C-4)

concentric: having a center in common. (p. 135, 11-D-7)

cone: a figure from geometry with a circular base and sides that taper to a point. (p. 109, 10-B-5)

constant: a quantity that always has the same value. (p. 57, 6-B-3)

contaminated: poisoned; having dirt in it. (p. 159, 12-D-3)

continuous: going along without a break; unbroken; connected. (p. 19, 3-A-4)

contour: the outline or shape of something. (p. 95, 9-A-2)

controls: an instrument used to direct the operation of a machine. (p. 104, 10-A-9)

conventional: doing something in a way that most people do. (p. 4, 1-B-6)

convex: curving outward, like the outside surface of half a rubber ball. (p. 131, 11-C-5)

coolant: a cutting fluid used to cool the tool and workpiece, especially in grinding operations; it's usually water based. (p. 99, 9-B-2)

coordinate: any number on an axis, used to tell the position of a point. (p. 195, 15-C-2)

coordination: muscles working together to produce accurate movements. (p. 159, 12-D-9)

core: the most important ideas and skills in an education program. (p.1, 1-A-10)

curled: twisted, bent around in the shape of a curl. (p. 19, 3-A-6)

curve: a line having no straight part or angled part. (p. 44, 5-B-10)

cut: a tear in the skin made by something sharp. (p. 159, 12-D-4)

cutting fluid: any of several materials used in cutting metals: cutting oils, synthetics, soluble or emulsified oils (water based) and sulfurized oils. (p. 99, 9-B-1)

cutting lips: the sharp edges of a twist drill which continue the cutting action begun by the chisel edge. (p. 111, 10-C-6)

cylinder: a shape which has straight sides and circular ends. (p. 19, 3-A-3)

D

dash: a short, thin line. (p. 32, 4-B-2)
degree: a unit of measure for angles and arcs. (p. 57, 6-B-1)
dent: a small hollow made in a surface by hitting it. (p. 85, 8-A-8)
detail: an important part of a part. (p. 32, 4-B-7)
development: the process of changing an invention to make it better. (p. 41, 5-A-2)
device: a mechanical invention made for some purpose. (p. 41, 5-A-3)
diagonal: a straight line running from one opposite corner to the other. (p. 41, 5-A-8)
dial: face of a meter, gage, or indicator on which a pointer indicates an amount. (p. 163, 13-A-7)
diameter: a straight line passing through the center of a circle from side to side. (p. 19, 3-A-9)
dimension: any measured length shown on a drawing. (p. 27, 4-A-7)
dovetail: an angled part that slides into a slot of the same shape. (p. 146, 12-B-6)
drive: a device that communicates motion to another part. (p. 44, 5-B-3)
drum: a cylindrical mount for tools; a tool can be rotated into position. (p. 191, 15-B-4)
durability: quality of lasting a long time, even with hard, frequent use. (p. 190, 15-A-8)

E

eccentric: having a center different from the common center. (p. 135, 11-D-8)
economical: able to be done without waste of money or time. (p. 173, 14-A-5)
edge: the thin cutting part of a blade. (p. 111, 10-C-1)
efficiently: doing something well without waste, delay, or extra expense. (p. 4, 1-B-10)
entry-level job: a first job that requires less skill, but is a way of starting a career. (p. 4, 1-B-3)
equivalent: equal in quantity, value, or meaning. (p. 51, 6-A-9)
exact: accurate without varying from a given number. (p. 38, 4-C-1)
execution: doing; acting on a plan. (p. 189, 15-A-5)
exploded view: a drawing showing the relationships of parts in an assembly. (p. 27, 4-A-5)

F

fastener: any part which is made to hold other parts together. (p. 19, 3-A-1)
ferrous: containing iron. (p. 173, 14-A-9)
field: any general area of knowledge or work which requires special training. (p. 4, 1-B-5)
fillet: a curved surface joining two intersecting surfaces. (p. 32, 4-B-4)
fine: thin, slender, or small. (p. 23, 3-B-5)
fire extinquisher: a tool used to put out fire by water or chemicals. (p. 12, 2-C-2)
fitting: a part used to join or adapt other parts, as in a system of pipes. (p. 181, 14-D-4)
fixed: not moving, remaining in one place. (p. 76, 7-B-6)
flammable: able to burn. (p. 12, 2-C-1)
flange: a rim which sticks out from a wheel to hold it in place and guide it. (p. 179, 14-C-4)
floating: not fixed; changing locations from one place to another. (p. 197, 15-D-2)
flute: the groove in a drill, tap, reamer, or milling cutter. (p. 44, 5-B-7)
focus: the center of attention; the main effort or idea. (p. 181, 14-D-9)
formula: a set of letters and numbers which tell about a mathematical relationship. (p. 51, 6-A-2)
foundation: the base on which something rests. (p. 181, 14-D-10)
frequent: happening often. (p. 51, 6-A-5)
friable: easily fractured into tiny pieces or easily crushed into powder. (p. 173, 14-A-4)
function: a quantity whose value depends on another quantity. (p. 57, 6-B-9)

G

gage (sawing): width of a saw blade. Gage is not as wide as tooth width (set). (p. 100, 9-B-9)
glazed: having a hard, glassy finish, as a grinding wheel with a loaded surface. (p. 179, 14-C-1)
graduation: a mark stamped on a measuring device to show smaller units. (p. 69, 7-A-9)
granite: a hard, gray or pink rock from which some surface plates are made. (p. 85, 8-A-5)
grit size: the size of the abrasive grains in the grinding wheel. (p. 176, 14-B-1)
groove: a long, narrow hollow cut into the surface of a workpiece. (p. 131, 11-C-8)
guard: a piece of metal which covers moving parts so workers will be safe. (p. 9, 2-B-6)
gullet: the bottom of the space between teeth on saws and cutters. (p. 99, 9-B-5)

H

handicapped: being limited in what you can do because of permanent injury. (p. 7, 2-A-6)
hazard: a danger. (p. 7, 2-A-2)
heading: a name written at the top of the page or paragraph to tell what is below. (p. 23, 3-B-8)
helix: unbroken line around and down a cylinder or cone, like a screw thread. (p. 111, 10-C-5)
high speed steel: a hard metal made from iron, carbon, and other chemicals. (p. 126, 11-B-2)
hinge: a device that joins two parts, but leaves them free to turn or swing. (p. 95, 9-A-6)
hollow: having an empty space inside. (p. 138, 11-E-3)
horseplay: touching a person in a playful, sometimes rough manner. (p. 13, 2-C-5)
housekeeping: keeping your tools and materials in order or put away. (p. 12, 2-C-3)
housing: a frame or box for holding and covering some mechanical parts. (p. 163, 13-A-2)
hydraulic: moved by the pressure of a liquid forced through an opening or tube. (p. 95, 9-A-9)

I

incremental: gradually increasing or decreasing by small amounts. (p. 197, 15-D-3)
independent: operating separately, without needing something else to happen. (p. 135, 11-D-9)
index: a pointer or indicator. (p. 76, 7-B-4)
injury: getting hurt. (p. 7, 2-A-4)
instrument: a carefully made tool that can be used for fine measurements. (p. 69, 7-A-4)
interchangeable: made the same, as a part that can be used anywhere in place of another. (p. 69, 7-A-2)
interface: the place where two things meet or touch. (p. 163, 13-A-3)
international: involving many nations and their relations. (p. 69, 7-A-3)
intricate: containing many unusual twists and turns of shape. (p. 95, 9-A-3)
invention: a mechanical product planned and made for a particular purpose. (p. 41, 5-A-1)
irregular: not regular; not having the same shape all the way around. (p. 135, 11-D-10)

J

jagged: having sharp, uneven points. (p. 126, 11-B-8)
joint: a part where two things are fastened together. (p. 41, 5-A-7)
juncture: a point, line, or area where things join or connect. (p. 131, 11-C-6)

K

keen: having a sharp edge or point that will cut well. (p. 126, 11-B-7)
kerf: the width of a cut produced by a saw. (p. 99, 9-B-4)
key pad: the rows of keys on a calculator. (p. 57, 6-B-10)
keyboard: a set of keys mounted on a board; it allows information and instructions to be typed into a computer. (p. 191, 15-B-2)
keyseat (or keyway): an axially located retangular groove in a shaft or hub. (p. 153, 12-C-2)

L

layer: a single thickness of some material. (p. 85, 8-A-7)
layout dye: a quick-drying, blue liquid which is applied to a workpiece surface. (p. 85, 8-A-6)
loaded: said of a wheel's cutting surface filled with tiny chips of metal. (p. 179, 14-C-2)
loose: not tight; having some space between parts. (p. 23, 3-B-7)
lowest terms: result of a fraction reduced until it can be reduced no further. (p. 51, 6-A-10)

M

machining center: a computer-controlled machine which has enough tools and positioning capabilities to perform several operations. (p. 197, 15-D-1)
machinist: a worker who uses machines to make parts out of materials like metal. (p.1, 1-A-1)
magazine: a place for storing things until they are needed. (p. 191, 15-B-3)
magnetic: ability of a piece of iron to attract other pieces made of iron. (p. 181, 14-D-7)
manual: done by the use of the hand. (p. 163, 13-A-5)
mathematical operations: any activity like adding, subtracting, multiplying and dividing which can be applied to quantities. (p. 51, 6-A-8)
mating parts: two parts which fit together. (p. 19, 3-A-7)
maximum: the highest amount possible. (p. 27, 4-A-9)
microinch: one millionth of an inch. (p. 38, 4-C-8)
micrometer (mike): instrument for measuring lengths, with high discrimination. (p. 76, 7-B-1)
Morse taper: a gradual decrease in width, thickness, or cross-section diameter by 5/8 inch for each foot of length. (p. 135, 11-D-3)
movable: able to be moved. (p. 76, 7-B-7)

N

nomenclature: a set of names used in a particular field of work. (p. 19, 3-A-8)
nominal: agreeing in value with a given number. (p. 38, 4-C-2)
notch: a concave cut in the edge or surface of an object. (p. 135, 11-D-4)

O

offset: bent twice to put the scriber point in a different position. (p. 90, 8-B-1)
on edge: the position of a part that has been placed on its narrower side. (p. 90, 8-B-4)
operation: the use of a machine tool in any way in the making of a part. (p. 85, 8-A-2)
optical: related to equipment which uses light to make very fine measurements. (p. 85, 8-A-3)

P

pair: two things or two parts which are used together. (p. 195, 15-C-1)
parallel: running in the same direction in a plane and equally apart at every point. (p. 90, 8-B-2)
pedestal: a column which supports something above it. (p. 181, 14-D-1)
periphery: the perimeter or external boundary of a surface or body. (p. 153, 12-C-3)
permanent: unchanging. (p. 7, 2-A-5)
perspective: a drawing in which things look smaller farther into the picture. (p. 27, 4-A-8)
pilot: a part of a cutting tool that guides the tool into a previously drilled hole. (p. 109, 10-B-6)
pitch (sawing): in saw teeth, the number per inch. (p. 99, 9-B-8)
plane: a flat surface in which plane figures, like circles and squares, can be drawn. (p. 57, 6-B-2)
pocket: an open space cut into some surface of a part. (p. 143, 12-A-2)
portable: able to be carried easily from one place to another. (p. 41, 5-A-9)
possibility: an activity which can be done in the near future. (p. 119, 11-A-10)
potential: possible; capable of happening. (p. 9, 2-B-1)
practice: an action which is learned and repeated many times; a habit. (p. 7, 2-A-9)

press: any machine by which something is squeezed or stamped by pressure. (p. 103, 10-A-1)
previous: coming before in time or order. (p. 51, 6-A-4)
production: the act of making something like parts, especially in large quantities. (p. 4, 1-B-9)
program: a set of directions for making something, written in a language that a computer-controlled machine can understand and execute. (p. 189, 15-A-6)
protractor: an instrument, with a half-circle shape, to measure or draw angles. (p. 85, 8-A-10)
pulley: a small wheel with a grooved rim in which a belt runs. (p. 103, 10-A-7)
Pythagorean theorem: a formula with which the length of any side of a right triangle can be computed, if the other two sides are known. (p. 57, 6-B-5)

Q

quadrant: a quarter section of a circle. (p. 195, 15-C-4)
quantity: the measured amount of something, expressed in numbers. (p. 51, 6-A-7)

R

rake: a tool angle that provides a keenness to the cutting edge. (p. 99, 9-B-6)
range: the set of numbers which includes variations from an exact amount. (p. 163, 13-A-8)
ratchet: a device which allows tightening a mike to a snug fit and then continues to turn without adding pressure. (p. 76, 7-B-2)
ratio: a comparison of one quantity to another. (p. 57, 6-B-8)
recent: done just before the present time; new; modern. (p. 189, 15-A-1)
recess: a hollow place, or a space moved back from the surface. (p. 76, 7-B-3)
reciprocating: moving alternately back and forth. (p. 95, 9-A-5)
reference: surface of known flatness; point from which to measure locations (p. 85, 8-A-4)
relief: clearance; space behind a cutting edge to allow material to slide by. (p. 111, 10-C-7)
relief angle: an angle that provides cutting edge clearance for the cutting action. (p. 99, 9-B-7)
reservoir: a place where a liquid is collected and stored. (p. 163, 13-A-4)
resinoid: like resin; a liquid, from plants or trees, that can be hardened. (p. 176, 14-B-5)
rigidity: condition of being unmoving and firmly fixed. (p. 163, 13-A-6)
rim: the edge, often of something circular (p. 176, 14-B-6)
rough stock: a piece of unworked metal from which a part can be made. (p.1, 1-A-3)
roughing: in machining operations, the rapid removal of unwanted material on a workpiece, leaving a smaller amount for finishing. (p. 99, 9-B-3)
round stock: a piece of unworked metal that is shaped like a cylinder. (p. 19, 3-A-5)

S

safety: being free from danger or injury. (p. 7, 2-A-1)
scrap: a part that has a dimension made outside of tolerance. (p. 38, 4-C-5)
section: a part or division of a book, a chapter, or a lesson. (p. 51, 6-A-3)
segment: any one of several parts into which a line can be divided. (p. 32, 4-B-6)
sensitive: aware of how something feels. (p. 103, 10-A-6)
sequence: a continuous series of actions done in order: step one, step two, etc. (p. 189, 15-A-3)
series: a number of similar things coming one after another. (p. 23, 3-B-3)
set: a collection of tools which are like each other, but are of different sizes. (p. 44, 5-B-4)
set (sawing): width of a saw tooth. Set is wider than blade width (gage). (p. 100, 9-B-9)
shank: the part of a tool which is held in a toolholder or in the hand. (p. 44, 5-B-8)
shape: the external surface or exterior form of something. (p. 131, 11-C-1)
shoulder: the part of a cylindrical object formed by a face and a side. (p. 131, 11-C-7)
sleepy: the condition of someone who is about to go to sleep. (p. 159, 12-D-7)
slot: a narrow groove cut into a part. (p. 143, 12-A-3)
smooth: having an even surface with no roughness. (p. 38, 4-C-7)

socket: a hollow part into which something fits. (p. 44, 5-B-2)

sore: an opening in the skin which is usually infected and painful. (p. 159, 12-D-5)

spacing collar: a short steel tube which slides onto the arbor of a horizontal mill as a means of positioning a cutter on the arbor. (p. 167, 13-B-1)

spindle: rotating tube with center axis; tapered chucks or drills can be inserted. (p. 103, 10-A-8)

spring-loaded: having a spring behind an arm or spindle; the spring allows the arm or spindle to return to its original position when pressure is removed. (p. 90, 8-B-10)

square root: number that, when multiplied by itself, gives the original number. (p. 57, 6-B-7)

standard: something established as a basis of comparison in measuring anything. (p. 69, 7-A-1)

stationary: not moving, fixed in one place. (p. 119, 11-A-1)

steady: not varying; holding in the correct position. (p. 138, 11-E-5)

stroke: a movement forward of the hand holding a saw or a file. (p. 44, 5-B-9)

structure: how something is built; arrangement of parts within the whole. (p. 176, 14-B-3)

stud: a cylindrical piece coming up from a surface; it can be threaded. (p. 135, 11-D-5)

support: machine part to carry weight of other parts and hold them in place. (p. 163, 13-A-1)

swarf: tiny chips produced when a grinding wheel is used to remove metal. (p. 173, 14-A-3)

symbol: a sign or picture which means something beyond itself. (p. 27, 4-A-6)

T

T-nut: threaded, T-shaped nut that fits into T-slots on a machine tool table. (p. 153, 12-C-4)

T-slot: T-shaped slot in a machine tool table, used to hold T-nuts and studs for clamping setups. (p. 153, 12-C-5)

task: work given to a worker by another person; a job requiring hard work. (p. 126, 11-B-5)

tendency: a leaning toward acting in a certain way. (p. 167, 13-B-3)

tight: placed closely together with no space between parts. (p. 23, 3-B-6)

tip: the pointed, tapered, or rounded end of something long and slim. (p. 109, 10-B-7)

tired: the condition of someone who has used up energy and needs to rest. (p. 159, 12-D-6)

(VERBS)

to aim: to point in a particular direction. (p. 119, 11-A-9)

to align: to bring a variety of parts into a straight line. (p. 109, 10-B-4)

to alternate: to take turns; to do something every other time. (p. 32, 4-B-3)

to anneal: to heat metal and then slowly cool it, in order to decrease hardness. (p. 100, 9-B-10)

to bind: to get caught in a narrow place, so motion is limited or impossible. (p. 167, 13-B-6)

to bond: to hold pieces together with material like glue or solder. (p. 173, 14-A-2)

to calculate: to use mathematics to get an answer. (p. 4, 1-B-8)

to calibrate: to compare a measuring instrument to a known standard of measurement and adjust it to be the same as that standard. (p. 69, 7-A-7)

to carry out: to do what was planned. (p.1, 1-A-7)

to chamfer: to remove the sharp edge of a part by cutting it at an angle, usually 45°. (p. 143, 12-A-4)

to chatter: to vibrate rapidly; said of a workpiece, a machine or a tool. (p. 126, 11-B-9)

to classify: to arrange into groups according to some quality (p. 23, 3-B-9)

to coincide: to have the same position in space. (p. 76, 7-B-9)

to communicate: to exchange understanding by using words or pictures. (p. 27, 4-A-1)

to compute: to arrive at a value, number, or amount by working a math problem. (p. 51, 6-A-1)

to cut off: to cut through stock to get a piece of useable length. (p. 95, 9-A-1)

to deburr: to use a tool to remove sharp, metal burrs from a workpiece. (p. 109, 10-B-2)

to deflect: to make something go to one side; to change course. (p. 181, 14-D-8)

to discriminate: to divide a basic unit into smaller units which can be used to measure lengths with the accuracy required by the job. (p. 69, 7-A-6)

to drive: to transmit motion to a machine part or to a workpiece. (p. 135, 11-D-1)
to elevate: to lift something. (p. 90, 8-B-5)
to engage: to interlock with, to mesh. (p. 145, 12-B-3)
to expose: to be out in the open air, to leave uncovered. (p. 159, 12-D-2)
to extend: to make something longer. (p. 32, 4-B-5)
to feed: to move a workpiece into a cutter or a cutter into a workpiece. (p. 143, 12-A-1)
to flare: to curve outward like the rim of a bell. (p. 176, 14-B-7)
to force fit: to push a part into a hole which is slightly smaller than the part. (p. 41, 5-A-10)
to fuse: to unite by melting things together. (p. 173, 14-A-6)
to grip: to hold onto something tightly. (p. 41, 5-A-6)
to guide: to lead in a particular direction. (p. 109, 10-B-3)
to hog: to remove large amounts of metal from a workpiece, using heavy cuts. (p. 153, 12-C-1)
to insert: to put into. (p. 135, 11-D-2)
to interpret: to show in what way you understand something. (p. 27, 4-A-2)
to intersect: to cross each other. (p. 27, 4-A-4)
to lay out: to measure and mark a workpiece to show the part's features. (p. 85, 8-A-1)
to loosen: to unfasten, to untighten. (p. 41, 5-A-5)
to magnify: to make larger, so an object can be better seen. (p. 23, 3-B-2)
to make sure: to do something carefully; to take care that something is done. (p. 159, 12-D-1)
to mate: to fit one part into another that is shaped to receive it. (p. 109, 10-B-1)
to monitor: to check a machine's performance and make adjustments if needed. (p. 191, 15-B-6)
to mount: to put or fasten in the proper place. (p. 44, 5-B-5)
to network: to join two or more computers so they can communicate with each other and share programs and information. (p. 191, 15-B-8)
to part: to cut or break into two or more pieces. (p. 131, 11-C-3)
to perform: to bring to completion; to do a task. (p. 131, 11-C-2)
to pinch: to squeeze between two surfaces. (p. 9, 2-B-5)
to plot: to mark the location of a point by the use of coordinates. (p. 195, 15-C-5)
to position: to put something in a desired place. (p. 145, 12-B-2)
to present: to put forward for use. (p. 119, 11-A-4)
to program: to give a computer some commands for doing something. (p.1, 1-A-9)
to project: to stick out from a surface. (p. 138, 11-E-4)
to protect: to keep from injury. (p. 7, 2-A-10)
to reciprocate: to move back and forth. (p. 181, 14-D-6)
to reduce: to lessen in any way, as in size, length, or weight. (p. 131, 11-C-9)
to refer: to send something in a certain direction from a given point. (p. 197, 15-D-5)
to regulate: to control; to direct; to adjust something so it works accurately. (p. 189, 15-A-4)
to remove: to take away. (p.1, 1-A-2)
to represent: to be or act in place of something or someone else. (p. 38, 4-C-6)
to require: to need, to be necessary. (p. 23, 3-B-1)
to reverse: to turn backward or in the opposite direction. (p. 138, 11-E-2)
to ring: to tap a grinding wheel and listen for the clear sound of a wheel with no cracks. (p. 181, 14-D-5)
to rotate: to turn around a center point or axis; to spin; to revolve. (p. 9, 2-B-2)
to sag: to sink, bend, or curve from weight or pressure. (p. 138, 11-E-6)
to secure: to hold something tightly in place. (p. 9, 2-B-3)
to sharpen: to grind a cutting tool edge to make it thin and better able to cut. (p. 111, 10-C-2)
to shatter: to break a hard object into sharp, jagged pieces. (p. 167, 13-B-7)
to slide: to move in a direction while touching a smooth surface. (p. 119, 11-A-6)
to slit: to make a long, thin, straight cut in any material. (p. 167, 13-B-8)
to slope: to run in a direction that is not level. (p. 111, 10-C-4)
to specialize: to develop skills in one part of a career in order to do the job well. (p. 4, 1-B-4)

to square (1): to multiply a number by itself. (p. 57, 6-B-6)
to square (2): to make sure all right angles on a workpiece are as close to 90° as possible. (p. 167, 13-B-10)
to stack: to put several things on top of each other in an orderly way. (p. 90, 8-B-6)
to stagger: to arrange along a line, with a tooth on one side, then on the other. (p. 167, 13-B-4)
to stamp: to imprint a mark, design, letter or number into metal by hitting the metal forcibly with a form of the desired shape. (p. 69, 7-A-5)
to subdivide: to divide a unit into smaller units and then divide those units again. (p. 69, 7-A-8)
to support: to hold up the weight of something else; to keep from falling. (p. 145, 12-B-1)
to swivel: to turn freely while being joined together. (p. 145, 12-B-4)
to take precaution: to be careful. (p. 159, 12-D-10)
to tighten: to put something together closely and securely. (p. 41, 5-A-4)
to tilt: to move something so that it slopes. (p. 95, 9-A-7)
to transmit: to send from one place to another. (p. 119, 11-A-8)
to traverse: to go back and forth over something. (p. 146, 12-B-5)
to vary: to have things be different in some respect from each other. (p. 167, 13-B-9)
to visualize: to see something in the mind.(p. 27, 4-A-3)
to warn: to let someone know about possible danger. (p. 9, 2-B-4)
to wring: to slide one clean gage block over another, so they stick together. (p. 90, 8-B-7)

(End of Verbs)

tool geometry: the correct angles and dimensions which should be ground on cutting tools in order to make efficient cuts and avoid tool wear. (p. 126, 11-B-4)
toolbit: the cutting tool used with a lathe; it has a single cutting point. (p. 119, 11-A-2)
toolpath: a line along which a tool moves during a machining operation. (p. 189, 15-A-7)
trade: a kind of work that requires special skills. (p.1, 1-A-6)
traditional: using methods that have been passed on from earlier times. (p. 190, 15-A-9)
trough: a long, narrow, open container for liquid or other loose material. (p. 181, 14-D-2)
turret: a tool-holding device which is able to rotate in a circular path to present a variety of tools at the point of cutting. (p. 119, 11-A-3)
typical: having the usual qualities. (p. 90, 8-B-8)

U

ultra-sonic: related to vibrations that are of a higher number than those usually heard. (p. 190, 15-A-10)
uniform: having the same shape or appearance along the length of an object. (p. 32, 4-B-10)
unilateral: having one side or measurement in only one direction. (p. 38, 4-C-10)
universal: operating together, not independently. (p. 138, 11-E-1)
unrelated: not leading to anything else. (p. 4, 1-B-2)
upset: the condition of someone who is angry or has other strong emotions. (p. 159, 12-D-8)

V

vacuum: suction provided to remove chips and coolant from a cutting interface. (p. 181, 14-D-3)
variation: change in form from what is usually seen. (p. 135, 11-D-6)
velocity: speed; distance covered in a unit of time. (p. 95, 9-A-10)
vernier: a short graduated scale that slides along a longer graduated instrument and is used to indicate fractional parts of graduations. (p. 76, 7-B-10)
vertical: in an up-and-down direction. (p. 13, 2-C-6)
visible: able to be seen. (p. 32, 4-B-1)
vitrified: changed into a glass-like substance by a process of high-heat fusing. (p. 176, 14-B-4)

W

wavy: in the shape of a wave; curved, not straight. (p. 32, 4-B-8)

width (sawing): the distance from the flat edge to the tip of the tooth on the sawing edge, measured across the blade. (p. 99, 9-B-8)

workpiece: the part as it is being worked on by the machinist. (p.1, 1-A-4)

INDEX

Abbreviations, 61
Abrasive machining, 174
Abrasives, 173-176
    aluminum oxide, 175
    cubic boron nitride (CBN), 175
    diamond, 175
    silicon carbide, 175
Absolute positioning in CNC, 198
Accuracy in measurement, 71
Adapters, milling machines, 143, 144
    156, 168, 171
Adjustable wrench, 45
Allen wrenches, 45
Alloys, 173
Aluminum oxide, 175
Angle plate, 90-92
Angles,
    in right triangle, 58
    measurement of, 58
    right, 58
Angular measuring instruments,
    bevel protractor, 88
    sine bars and gage blocks, 91, 93
Annealing, 100
Arbor press, 42
Arbors, horizontal mill, 144, 164, 168
Arbor supports, 164, 166, 168
Assembly drawing, 28
Automatic tool changer (ATC), 192

Balancing of grinding wheels, 180
Base,
    on horizontal mill, 164
    on lathe, 121, 125
    on vertical mill, 146, 147
Bearings, 164
Bed,
    cylindrical grinder, 187
    lathe, 120, 121, 125
Bevel protractor (combination set), 88
Bilateral tolerance, 39
Blades, sawing machine, 95, 99, 101

Bolts, 20, 25
Bonding in grinding wheels, 176
Box-end wrenches, 45

Calculator, 57, 64
Calibration, 71
Calipers,
    dial, 83
    outside micrometer caliper, 77
    vernier, 81
Carbide tools, 171, 175
Career, 4, 5
    career opportunities, 5
Carriage, lathe, 121, 123
Cartesian coordinate system, 195, 196
CBN, *See* cubic boron nitride,
C-clamps, 42
Centerhead (combination set), 88
Center lines, 34
Center punch, 86, 87
Centers, turning between, 136, 137
Chamfer, 32
Chips, 1, 2
    safety in handling, 10, 11
Chucks,
    collet, 140
    drill, 103, 105, 109
    four-jaw independent, lathe, 137
    magnetic, grinder, 181, 183, 184
    three-jaw universal, lathe, 138
Circle,
    degrees in a, 58
    parts of, 59
Clamps, 42
Clearance, 116, 126, 127
CNC machine operator, 194
CNC programmer, 194
CNC set-up person, 194
Codes for CNC machining, 195
Collets, 140, 157
Comparators,
    dial indicator, 91, 93, 94

Compressed air, 15, 17
Computer numerical control (CNC),
    1, 2, 189-198
    advances in, 190
    automatic tool changers (ATC), 192
    incremental & absolute positioning,
        198
    machining centers, 192
    MCU input, 192
    programming for, 194
Concentricity, 135, 137
Controls 104
    for drill press, 105
    for grinding machines, 177, 184
    for lathes, 121, 123
    for milling machines, 165
Coolant, 164, 184
    and safety, 11
Coordinates, 195, 198
Cosine function (Cos) in trigonometry,
    63, 64
Counterboring, 100
Countersinking, 100
Crescent wrenches, 45
Crossfeed,
    on grinder, 183, 184
    on lathe, 121
    on mill, 148, 165
Cubic boron nitride (CBN), 175
Curved tooth files, 46
Cutoff machines, 97
    horizontal band saw, 98
    reciprocating saw, 98
    universal tilt frame, 98
Cutting speed formula, 61
Cutting tools,
    arbor-driven for horizontal mill,
        167-172
    for lathes, 127, 128, 132-134
    for vertical mills, 154-157
    safety in the use of, 160, 161, 169
    sharpening of, 182
Cylindrical grinders, 186-188
Decimal fractions, 54
Decimal inch, 72

Decimial inch ruler, 75
Depth gage, dial, 82
Depth micrometer, 77, 80
Detail drawing, 28
Diagonal cutters, 42
Dial indicator, 91, 93, 94
Dies,
    adjustable split, 48
    rethreading, 48
    thread cutting, 48
    two piece, 48
Dimensions, 34, 38, 39
Discrimination, 71
Dovetails, 146, 164
    dovetail ridge, 147
    dovetail slot, 147
Downfeed handwheel, 183-186
Drawings,
    dimensions, 34, 38
    exploded view, 29
    finish marks on, 40
    isometric, 29, 33
    orthographic projections, 29, 30
    perspective, 27, 29
    reading and interpreting, 28
    scale, 39
Dressing of grinding wheels, 179, 180
Drilling on the drill press, 105, 110
Drill presses, 105-108
    twist drills and, 105
    types of,
        radial arm, 106, 107
        sensitive, 105, 107
    work holding with, 106
Drill sizes, 117

Ear protection, 17
Electricity, safety with, 13, 14
End mills,
    ball-end, 155
    flycutters, 156
    number of flutes, 154
    shell, 156
    single angle, 155
    T-slot cutters, 155

Woodruff keyseat cutter, 155
Engine lathes, *See* lathes
Entry-level jobs, 5, 6
Equivalents of decimals for
    bar fractions, 73
Exploded view drawings, 29
Eye protection, 9, 10, 16

Facing, 132, 171
Feeds,
    CNC, 192, 197
    drills, 104, 105
    grinders, 183, 184, 187
    horizontal mill, 165
    lathe, 121, 123
    vertical mill, 143, 144, 148
Files, 46, 49
    curved tooth, 46
    half-round, 46
    parts of, 46
    pinning of, 46
    round, 46
    square, 46
    thread, 46
    triangular, 46
Finish marks on drawings, 40
Fire extinguishers, 13, 17, 18
Fits,
    force fits, 41
Footstock, 187
Formulas, 51
Fractional inch, 72
    dimensioning with, 72
    rules, 75
Fractions,
    bar and decimal, 53, 54
Friability, 173-175

Gage blocks, 90, 91, 92
    calculating stack combinations,
      90, 93
    use of wear blocks, 93
    wringing of, 90, 93
Gages,
    depth, 82

plug, 84
ring gage, 84
screw pitch, 84
small hole, 84
surface, 84
telescoping, 84
thread plug, 84
thread ring, 84
vernier height, 91, 92
General machinist, 5, 6
Glasses, safety, 9, 10
Goggles, safety, 9, 10
Grains, abrasive, 173-176
Grinders,
    cylindrical, 186-188
    chucks for, 181
    pedestal, 181, 182
    surface, 183-185
    universal center-type cylindrical, 187
Grinding,
    abrasives for, 175
    economical method, 174
    friability of grains, 174
    lathe toolbits, 127
    ring testing of wheels, 178, 182
    safety while using, 182
    selection of wheel, 175, 176
    size and shape of wheels, 178, 179
    with magnetic chuck, 181, 183, 184
Grinding wheels,
    balancing of, 179, 180
    oonds for, 176, 177
    truing and dressing, 179, 180

Hacksaw, 46
Half-round file, 46
Hammers, 86, 87
Hand-grinding on pedestal grinder, 182
Hand protection, 10, 11
Hand tools, nomenclature of, 49, 50
Hazards in the machine shop,
    clutter, 14
    fumes, 11
    heavy lifting, 12, 17
    horseplay, 15

noise in the ears, 17
objects in the eyes, 9
to the hands, 10
to the lungs, 11
Headstock, 121, 122
Helical flutes, 111, 113, 155
Hex bolts, 25
Hex nuts, 25
High speed steels (HSS), 126, 127
Horizontal band cutoff machine, 98, 100
Horizontal milling machines, 163-172
    adapters, 171
    arbors, 144, 164, 168
    arbor-driven milling cutters, 168-172
    coolant, 164
    cutters, selection of, 168-171
    knee and column, 164
    reservoir, 164
    speed change dial, 165
    universal, 164
    using face milling cutters, 171
Housekeeping, 14
HSS steels, *See* high speed steels

Incremental positioning in CNC
    programming, 198
Indicators, *See* comparators
Inside chucking, 139
Inside diameter, 61, 139
Inside micrometer, 76, 77
Interchangeable parts, 69, 70
Internal cylindrical grinder, 186, 187
International metric system, 72
Isometric drawings, 29, 33

Kerf, 99, 100
Keypad, 57, 64
Keys, 155
Keyseats, 155
Knee,
    horizontal milling machine, 164, 165
    vertical milling machine, 147, 148
Knee and column milling machines, 164

Lasers and precision, 87

Lathes,
    alignment of centers, 136, 137
    base, 121
    bed, 121
    carriage, 121
    cutting tools for, 132-133
    mandrels for, 140, 142
    steady rests for, 138, 139
    turning between centers, 132-137,
    turning to size, 133
Layout,
    dye, 86, 87, 89
    hammer and punches, 86, 87
    precision, 87
    scribers, 86, 87
    semi-precision, 87
    surface gage, 91, 93
    surface plate, 85-87
    tool nomenclature, 86, 91
Lifting, 12
Light, measurement with, 85
Lines, used in drawings, 33-37
    break, 35
    center, 34
    cutting plane, 36
    dimension, 34
    extension, 34
    leader, 35
    object, 33
    section, 36
    visible, 33

Machine control unit (MCU), 192, 194
Machine operator, 5, 6, 194 (CNC)
Machinist's combination set, 88
Mandrels, 140, 142
Mass production, 70
Math terminology test, 67-68
Measurement,
    accuracy, 71
    calibration, 71
    coordinates for, 195-198
    discrimination, 71
    principles (terms) of, 71
    reliability, 71

systems of,
    English (inch) system, 72
    metric system, 74
Metric/inch rules, 75
Metric measurement system, 74
Metrology, 71
Microinches, 40
Micrometers,
    caliper-type outside, 77
    depth, 77
    how to read, 78-79
    inside micrometer, 77
    outside micrometer, 69, 77
    vernier micrometers, 79
Milling cutters,
    horizontal, 168-172
    vertical, 154-157
Morse tapers, 105, 135, 136

Needle nose pliers, 42
Nominal size, 39
Nuts, 25

Open-end wrenches, 45
Orthographic projections, 30, 31, 33
Outline, how to make, 96-98
Outside chucking, 139
Outside diameter, 61, 139
Outside micrometers, 76-79
Overarm, horizontal mill, 164

Parallel, 90
Parallel bars, 91, 92
Pedestal grinders, 182
Phillips screwdriver, 45
Pi (π), 59
Pin punch, 86, 87
Pitch diameter, 21
Pitch, screw, 21
    screw pitch gage, 84
Pliers, 42
Power hacksaw, 98
Precision in measurement, 71
Program, 1
Punches, 86, 87

Pythagorean theorem, 57, 61, 62

Quick-change gearbox, 121, 123
Quick-change tooling, 121 122
Quill, 144, 146, 149

Radial arm drill press, 106-108
Rake on cutting tools, 101, 127-129
Ram-type vertical milling machine,
    See vertical milling machine
Ratios, trigonometric, 63
Reamers and reaming, 47, 110
Reliability in measurement, 71
Right angle, 58
Right triangle, 58
Rough stock, 1, 2
Round files, 46
Rules, *See* steel rules

Saddle,
    horizontal mill, 164, 165
    lathe, 121, 123
    vertical mill, 146, 148
Safety,
    attitude of, 8
    carrying heavy objects, 12, 17
    carrying long objects, 15, 17
    chip handling, 11
    clutter and housekeeping, 14
    compressed air, 15, 17
    eye protection, 9, 10
    fire extinguishers, 13, 17, 18
    gloves, 10
    hair, 15
    horseplay, 15
    jewelry, 15
    respiration, 11
    short sleeves, 15
    ventilation, 11
Saw blades,
    arbor-driven slitting saw, 170
    band saw blades, 98
    blade nomenclature, 99, 101
Sawing machines,
    abrasive cutoff machine, 98

cutting fluids, 99
gage of sawing blade, 100, 101
horizontal band cutoff machine, 98
reciprocating machines, 98
saw teeth, 99
set of sawing blade, 100
Scale on drawings, 39
Screwdrivers,
  phillips, 45
  standard, 45
Screws, 20, 25
Scribers, 86, 87, 91, 92, 93
Sensitive drill press, 105
  nomenclature of, 107
Setup person (CNC), 194
Silicon carbide, 175
Sine function (sin) in
  trigonometry, 63, 64
Sine bars, 91, 93
Slide and swivel tables, 187
Slip-joint pliers, 42
Small hole gages, 84
Sockets, 45
Socket wrenches, 45
Spacing collars, 168
Spindle,
  cylindrical grinder, 186
  drill press, 105-107
  horizontal mill, 165
  lathe headstock, 121, 122
  vertical mill, 146, 149
Square root of a number, 61
Squaring a number, 61
Standard screwdriver, 45
Steady rests for lathes, 138, 139
Steel rules, 75
  reading decimal inch rules, 72, 75
  reading fractional inch rules, 72, 75
Stud bolts, 19
Surface finish, 40
Surface gages, 91, 93
Surface grinders, 183-185
Surface plate, 85-87
Surface roughness, 40
Swarf, 174

Symbols on drawings, 27, 40

Tables
  cylindrical grinder 187
    swivel, 187
  drill press, 105, 106
  horizontal mill, 164
    table swivel, 164
  sawing, 98, 100
  surface grinder, 183, 184
  vertical mill, 148
Table feed handwheel, 148
Tailstock, 121, 123
Tangent function (tan) in
  trigonometry, 64
Taper,
  Morse taper, 105, 135, 136
  standard national milling machine
    taper, 168
Tapping,
  by hand, 47
  with drill press, 100
Taps, 47
Tap wrench, 47
T-bolts, T-nuts, T-slots, 153, 164
Telescoping gages, 84
Threads,
  class of fit, 24
  size of, 24, 25
Thread form,
  parts of, See Unified thread form
Tolerance,
  and gage blocks, 93
  bilateral, 39
  defined, 39
  nominal size and, 39
  unilateral, 39
Toolbit nomenclature,
  127, 128, 130
Toolhead, vertical mill, 146, 149
Toolholders, 121-123
Toolposts, 121-123
Triangles, 58
Triangular file, 46
Trigonometric functions, 63-66

Trigonometry, definition of, 63
Truing grinding wheels, 179, 180
Turning machines, *See* Lathes
Turrets, 120, 149, 192
Twist drills, 112-116

Unified thread form, nomenclature for,
    21, 22, 26
Unilateral tolerance, 39
Universal milling machine, 164

Vernier calipers, 81
Vernier devices, 77, 81
Vernier height gage, 91, 92
Vernier micrometer, 77
Vernier scales, 82
Vertical band saw, 97
Vertical milling machine, 143-162
    base and column, 146, 147
    controls, 146, 147, 148
    knee, 146, 147
    quill, 146, 149
    ram, 146, 149
    saddle, 146, 148
    shank cutters, 144
    spindle, 146, 149
    table, 146, 148
    toolhead, 146, 149
    turret, 146, 149
    vertical axis, 144
Vise, bench, 42

Ways, 121, 147
Welding, of band saw blades, 100
Wheelhead, for grinding machine, 184
Woodruff keys, 155
Woodruff keyseat cutter, 155
Workpiece, 1, 2
Wrenches, 45
Wringing, gage blocks, 90, 93